Climate Resilient Sustainable Agriculture

Kodoth Prabhakaran Nair

Climate Resilient Sustainable Agriculture

Springer

Kodoth Prabhakaran Nair
Kozhikode, Kerala, India

ISBN 978-3-031-84165-1 ISBN 978-3-031-84166-8 (eBook)
https://doi.org/10.1007/978-3-031-84166-8

This Springer imprint is published by the registered company Springer Nature Switzerland AG
The registered company address is: Gewerbestrasse 11, 6330 Cham, Switzerland

I dedicate this important book to the memory of Late Mrs. Mavila Chandravathy Nambiar, eldest sister of my wife, Dr. Mrs. Pankajam Nair. Her empathy and benevolence were her sterling qualities. She was a pillar of great strength to many in our family. May her soul rest in peace.

Preface

The Nefarious Activities in Ecologically Sensitive Areas

World Weather Attribution, a global consortium of climate scientists who study the impact of global warming on extreme weather events, also highlighted in its August 14, 2024, report that "factors such as quarrying for building materials" have definitely contributed to the increased susceptibility of the slopes to landslides during heavy rainfall. It is also important to note that illegal quarrying generates a lot of black money—a perennial financial menace in Kerala State.

To understand the spread of quarries in the Wayanad district, which ranks 13th of the 147 landslide vulnerable districts in India, as per the Landslide Atlas of India, the Satellite imagery of quarrying sites from open-source database OpenStreet Map was analyzed. The imagery was then superimposed on a map of the district to assess if the quarries fall in ecologically sensitive areas (ESAS) identified by the High Level Working Group (HLWG) on the subject in 2013.

HLWG was established by the Union environment ministry in April 2012 to formulate an "all-round and holistic approach to sustainable and equitable development" for the Western Ghats. This Group was led by Dr. Kasturirangan, former chairman of the Indian Space Research Organization (ISRO). The group identified 4156 villages, spanning 59,940 km^2—37 per cent of the mountain range—as "ecologically sensitive" and recommended a strict ban on quarrying, in these areas. These villages had at least 20 per cent of the area under natural landscape, including regions with high biological diversity, wildlife corridors or heritage sites. Accepting the recommendations, the Union environment ministry in November 2013 issued directives under section 5 of the Environment Protection Act, 1986, which prohibits quarrying, mining and other environmentally damaging activities in ESAS. The very surprising and distressing fact is that 15 of the 48 quarries in Wayanad district are located in the 13 villages identified as ESAS. In fact, 9 of the quarries are within forests in two ESA villages, namely, Kidanganad and Noolpuzha.

Explosives' use in quarrying generates vibrations which disturb the soil profile, which is particularly dangerous for Wayanad's critical slopes prone to landslides.

Quarrying not only creates fractures in the rock but also widens existing ones, allowing water to very speedily seep through following heavy rainfall, which ultimately, leads to landslides, due to the loosened soil profiles.

A Protracted Battle

Even after a decade of establishing a legal framework to ban mining and quarrying in the Western Ghats, India is yet to finalize the boundaries of ecologically sensitive areas to implement it. The following is a chronological assessment of the various initiatives:

1. August 31, 2011: The Western Ghats Ecology Panel submits a report to protect Western Ghats. Recommend strict restrictions that are opposed by the Kerala Government
2. October 29, 2013: Kerala Government sets up a three-member panel expert committee to study citizens' concerns on the HLWG report
3. December 2013: Centre asks the 6 states of the eastern Ghats to suggest modifications in the boundary of the ESA villages
4. March 10, 2014: Centre issues first draft notification to finalize ESA boundaries, but, it lapses as states demand smaller area under ESA. More draft modifications released in 2015, 2017, 2018, and 2022, but, all lapsed
5. August 17, 2012: The center sets up HLWG to find a balance between conservation and sustainable development. Report submitted on April 15, 2013
6. November 13, 2013: Centre accepts HLWG report and issues directives under section 5 of the Environment (Protection) Act, 1986. Prohibits quarrying in villages and ESA as identified by the HLWG report
7. December 3, 2018: Centre accepts Kerala Government's demands and amends EPA, brings down ESA area spread from 13, 108 km^2 to 9993 km^2
8. July 31, 2024: Centre releases draft notification after the Wayanad landslides to finalize the ESA boundaries!

What Is in Store for the Future?

More Wayanad—like disasters—could be in the making across Western Ghats as none of the states that shares the mountain range have finalized the ESAS even though a decade has passed since the Union Environment Ministry accepted the Kasturirangan report and amended the ESA to provide a legal framework.

Kerala currently has 110 quarry permits awarded up to 12 years and 444 quarry leases awarded up to 3 years according to the Department of Mining and Geology. 11 of the 14 districts in Kerala have ESAS as identified by Kasturirangan report. There are now 1688 quarry sites across Kerala, with 90 of them spread across 123 ESA villages. These quarries exist at a time when the state has not only weakened its regulations, but has miserably failed to implement its mining laws. A 2017 Comptroller and Auditor General (CAG) report found that five Kerala districts, including Wayanad, lacked a system to ensure compliance by quarries. Of the 27 quarries inspected, 21 violated rules. The July 30 landslides simply proves that there is blood on the LDF government's hands.

It is against the above-described background that this book has been written, so that the world knows how to preserve the fragile lands for agriculture. It is the hope of this author that this book will be guiding principle to embark upon climate resilient agriculture.

Kozhikode, India Kodoth Prabhakaran Nair

Acknowledgements

I very gratefully acknowledge the conscientious support of Ms. Margartet Deignan, Executive Editor, Springer, who, despite her grave personal bereavement, lent me a helping hand in the speedy publication of this important book. I owe her a deep debt of gratitude. Mrs. Aarthi Padmanabhan, Production Editor, Springer, and Mr. Madanagopal, Production Supervisor, and their team did a marvelous job in executing the production of this important book. My sincere appreciation to them.

Contents

Chapter 1
Weather and Climate

Weather and climate have profound influence on agriculture in a region. There are some important differences between the two terms. Weather describes the state of the lower atmosphere of a place with respect to temperature, annual precipitation, wind direction, wind speed, humidity, cloudiness and pressure etc., over a short period of time. The short term variations over minutes to days are indicated as weather of a location. It says whether the atmosphere is cold or hot, rainy or dry, sunny or cloudy, still or cyclonic, and, so on, usually for a period of 24 h. The weather may vary from place to place and from time to time, due to difference in temperature, cloudiness, humidity, rain, and, other atmospheric events. It may be sunny at one locality, but, rainy in another. Similarly, the weather of a place in winter morning may be foggy and sunny after a few hours as the fog disappears. Variation in weather is driven by Sun as it heats up air in the lower atmosphere at varying degrees. Meteorology is the study of weather meteorologists predict/forecast the weather of a station based on the observations recorded I the station or date collected from satellites. Weather prediction is done by analyzing the meteorological data from satellites and weather stations, which are very important, as weather influences many human activities. Important activities influenced by weather are, agriculture, sailing, fishing, AI flight, fire fight, trading, and, military operations.

The statistical information which indicates variations of weather at a given place for a specific interval of time is defined as climate. Climate usually describes the atmospheric changes over longer periods of time, normally three decades or more, as defined by the World Meteorological Organization (WMO), located in Geneva, Switzerland. WMO is the United Nations system's authoritative voice on the state and behaviour of the Earth's atmosphere, its interaction with the land and oceans, the weather and climate it produces and the resulting distribution of water resources. WMO is the United Nations system's authoritative voice on the state and behaviour of the Earth's atmosphere, its interaction with the land and oceans, the weather and climate it produces and the resulting distribution of water resources.

K. P. Nair, *Climate Resilient Sustainable Agriculture*,
https://doi.org/10.1007/978-3-031-84166-8_1

Table 1.1 Differences between weather and climate

Trait	Climate	Weather
Definition	Climate refers to statistical weather information that indicates the variation of weather at a given place for a specified interval	Weather is the day-to-day state of the atmospheric situation and it's short-term variation
Components	Climate includes precipitation, temperature, humidity, sunshine wind velocity, fog, frost and hail storms over a long period of time	Weather includes temperature, sunshine, rain, cloud cover, winds, hail, snow, sleet freezing rain, flooding, ice storms, thunder storms, heat waves etc., as short-term events
Forecast and time	By aggregates of weather statistics over 30-years	By collecting meteorological data, such as, air temperature, pressure, humidity solar radiation, wind speed, and its direction over a period of 24 h
Investigation	Climatology	Meteorology

Usually, climate is understood as the long-term weather pattern in a particular area. The atmospheric events like a hot/dry spell or cold spell indicates the weather of the place m but, not its climate.

Both weather and climate have some common components, such as, rainfall/precipitation, wind speed and direction, its velocity, humidity, atmospheric pressure and temperature. Climate is considered as average weather conditions of a region over time and space.

Differences Between the Terms "Weather" and "Climate"

The above-mentioned terms are used interchangeably, sometimes. But, they differ in terms of the length of time they measure and what trends affect them. These two terms are usually summarized by the popular phrase "Climate is what you expect and weather is what you get". The scientists describe climate as averages of precipitation/rainfall, temperature, humidity, sunshine, wind velocity, phenomena such as, fog, frost, hailstorms and other measures of the weather over a period of three decades. The differences between both the terms ae explained in Table 1.1.

What Are the Elements of Climate and Weather?

While many elements make up the weather of a region, temperature, atmospheric pressure, wind speed and direction, humidity, precipitation/rainfall, visibility, cloudiness, solar radiation, are the most vital. These are the primary elements of climate and weather.

Temperature: It refers to the amount of heat energy experienced in the air which is measured in centigrade degrees (Celsius) or Fahrenheit degrees. Celsius and Fahrenheit are two scales used to measure temperature. The temperature in the centigrade scale will be expressed in degrees Celsius. The temperature in the Fahrenheit scale will be expressed in degrees Fahrenheit. The relation between Celsius and Fahrenheit is proportional. Temperature variations during day and seasons are caused due to the Earth's rotation and transitional motion of the Earth around the Sun.

Atmospheric pressure: Atmospheric pressure is created by weight of air on a given surface of the Earth's atmosphere. The air around you has weight, and it presses against everything it touches. That pressure is called atmospheric pressure, or air pressure. It is the force exerted on a surface by the air above it as gravity pulls it to Earth. Atmospheric pressure is commonly measured with a barometer. The atmospheric pressure varies vertically and the vales decreases as we go up in altitude. As elevation increases, there is less overlying atmospheric mass, which causes a decreased pressure with increasing altitudes.

Wind speed and direction: The movement of air (wind) is caused by the large scale movement of air from an area with high pressure to an area with low pressure. The extreme weather events like cold and warm fronts, clouds, thunderstorms, and hurricanes are driven by the wind. There is a close relationship between pressure gradient and circulation of air. Control of air motion is regulated by the pressure gradient. The force which deflects the direction of the wind is called "Coriolis Force". The Earth rotates from west to east and completes one rotation in 24 h. Coriolis force is an apparent force caused by the earth's rotation. The Coriolis force is responsible for deflecting winds towards the right in the northern hemisphere and towards the left in the southern hemisphere. This is also known as 'Ferrel's Law'. The deflection is more when the wind velocity is high.

Humidity: At a given time when the amount of water vapor in the atmosphere is measured it refers to humidity. Sine water is formed after evaporation of water, it is not visible. Humidity plays a huge role in weather formation. It is measured by an apparatus called "Hygrometer". And this measure is expressed as absolute humidity, relative humidity and specific humidity. Relative humidity, often expressed as a percentage, indicates a present state of absolute humidity relative to a maximum humidity given the same temperature. Specific humidity is the ratio of water vapor mass to total moist air parcel mass. Humidity plays an important role for surface life.

Precipitation: Precipitation takes places when water vapor condenses and falls to the ground after it becomes too heavy to stay suspended in the air. Dust particles or smoke in the atmosphere are called "condensation nuclei", which are required for precipitation. Precipitation includes snow, hail, sleet, drizzle, fog, mist and rain. A large percentage of runoff water from watersheds drains into rivers, lakes and ocean, while the rest evaporates or passes through plants. The raindrops may be smaller (1 mm), larger (2–3 mm) or very large (4.5 mm). While small drops are spherical, larger drops may be like kidney beans and very large drops may look like parachute. Of all forms of precipitation rainfall is a major source which is measured by a rain gauge installed in a rainfall recording station. Precipitation is always considered as fresh water, even when the water originated from the ocean. However, pollutants in the atmosphere may contaminate water droplets before they fall to the ground. Acid rain is often caused due to such pollutants. The rainfall pattern, involving quantity, quality, and distribution are very important for the crops grown in rain-fed areas.

Visibility: This is a critical element of weather which measures the degree through which an object can be observed from a certain distance. Visibility is obstructed by cloud, mist, haze and fog, as well as, precipitation. It is represented by the meteorological optical range (MOR) and sensors like "forward scatter sensor" are used to measure visibility. In extremely clean air, the visibility can be up to 70–100 km. Due to air pollution, and high humidity, visibility is reduced. During sandstorm in desert visibility is often reduced. Blizzards (blowing snow) also causes a reduction in visibility.

Cloudiness: These are suspended molecules of war in different states that are formed in the atmosphere when humid air cools down to its dew point, and, water droplets or ice attach to small particles of dust, ash or other contaminants. Clouds indicate rain. Clouds are classified as low clouds, average clods and high clouds.

Low clouds: Strains, nimbostratus, stratocumulus, cumulus, towering cumulus and cumulonimbus.

Average clouds (6500–20,000 ft.): Altostratus, altocumulus and lenticular.

High clouds (over 20,000 ft.): Cirrus, cirrocumulus and cirrostratus.

These clouds appear in different forms, such as, the following:

1. Cirro—curl of hair, high
2. Alto—mid
3. Strato—layer
4. Nimbo—rain, precipitation
5. Cumulo—heap.

Clouds play multiple roles in climate system. Satellite and radar images are used to measure cloud density, the amount of moisture, the temperature and movement of clouds.

Clouds are the base for precipitation and they effectively reflect light to space and thus contribute to atmospheric cooling. During summer, clouds provide protection from the severe sun rays and in winter, they reduce nocturnal radiation and check the fall of temperature. A small increase in cloud cover can balance the heating resulting from greenhouse gases (GHGs).

Solar Radiation

Sunshine duration is the length of time the Earth's surface is directly exposed to solar radiation. It is also referred to as brilliant sunshine hour (BSH). The sunshine caused by solar radiation greatly influences other elements of the weather, such as, ambient temperature, humidity and air pressure, which can change the whole make up of the weather conditions. The instruments used to record sunshine is known as Campbell–Stokes recorder.

Solar radiation comes in many forms, such as, visible light, radio waves, heat (infrared), X-rays, and ultraviolet rays. Solar radiation is higher on clear, sunny days and usually low on cloudy days.

Climate and season: While climate is the average weather conditions, in a certain area over a long period of time, the seasons are the changes in the climate throughout the year in terms of factors, such as, latitude, altitude, and proximity to water. The climate of a place can be called as, temperate, tropical, hot or warm summer, and cool winter, whereas tropical climate has hot summer and warm winter. Polar climate is experienced with cold winter and cool summer. Seasons are caused by the Earth's tilt on its axis as the sun's rays hit the Earth at different angles, throughout the year. The main seasons are spring, summer, fall and winter. In India the main seasons are rainy (wet), winter and summer (dry) with intermediary seasons like autumn pre-winter and spring.

Climate Classification

To understand scientifically, climate classification has been made of different regions. Modern classification of climates has been started from the nineteenth century onward using empirical and generic methods. But, two popular methods, known as, Koppen and Thornthwaite methods are invariably used for classification of climate, which are explained below.

Wladimir Koppen (1846–1940) proposed the empirical classification and several revisions, thereafter. He used certain critical values of temperatures of the warmest and the coldest months and rainfall of the wettest and the driest months coinciding with the vegetation divisions. As assumed by Koppen, vegetation was used to represent major ecosystems and biomes on the Earth. Koppen considered mean monthly and annual temperature and mean monthly and annual precipitation for delineation of climatic regions. He classified the climate of the world into five climatic zones, based on temperature, which allowed different growth of vegetation, as described below. Koppen's map showed different colors and shades to represent the different climatic zones of the world.

Zone A: This covers tropical and equatorial zone shown in blue color.

Zone B: This covers arid dry zone indicated in red, pink and orange colors.

Zone C: This covers warm/mild temperate zone indicated in green color.

Zone D: This covers continental zone indicated in purple, violet, and light blue color.

Zone E: This covers the polar zone indicated by grey color.

These zones, described above, are known as tropical, dry, mild, cold, temperate and polar zones

Tropical (A): In tropical climate, the temperature in cooler months does not exceed 18 °C. Annual precipitation is more than 1500 mm.

Dry climate (B): In this zone, the precipitation is little and no month exceeds an average temperature of 10 °C. Perpetual water deficiency is experienced.

Humid mesothermal (C): Warm and humid summer with mean temperature of coldest month being below 18 °C but above − 3 °C. And, the warmest month is above 10 °C.

Humid microthermal or cold forest climate (D): Warm to cool summer with the coldest month temperature below − 3 °C and warmest month temperature above 10 °C.

Polar climate (E): Year-round cold with temperature of every month below 10 °C and without a warm season.

Koppen used English capital letters (Am B, C, D, E, F, M, S, T, W) and small letters (a, b, c, d, f, h, k, m, s, w) for climatic classification. The capital letters indicated the characteristic of vegetation and small letters indicated the precipitation and heat. Koppen's scheme uses letter symbols to denote various characteristics, which is practical and convenient. However, it ignores other factors, such as, wind, cloudiness, rainfall intensity, currents and, above all else, the air mases which for the basis of modern climatology. It is also difficult to explain the existence of different vegetation types within the same climatic division and similar vegetation types in different climatic divisions.

The sub-divisions with characteristic of these five groups, explained above, are described in Table 1.2.

The Thornthwaite Climate Classification

The Thornthwaite (1948) climate classification is based on moisture and heat as primary factors and seasonality as secondary factors in 1931 which was further modified in 1945. It is an improved and modified version of Koppen's climate classification. Thornthwaite assumed that there was a correlation between precipitation, temperature and plant association. The growth and development of natural vegetation depended on precipitation, rate of evaporation and thermal conditions of a region. The classification of climates was based on the concept of potential evapotranspiration (PET), which is equal to the total amount of water lost by way of evaporation and

Table 1.2 Koppen's climate classification

Group	Type	Region	Characteristics
A. Tropical	Af: Tropical rainforest	Amazon basin. Zaire south-east Asia, Congo, Singapore, Hawaii	Temperature exceeds 18, mean temperature 20 average rainfall 250 cm, luxurious vegetation
	Am: Tropical monsoon	Ivory Coast, Indonesia, Africa, Australia	Seasonal wind reversal
	As: Dry summer	India (Chennai, Odisha)	Winter rains
	Aw: Tropical savanna	Mumbai, Ho Chi Minh	Dry winter
B. Dry (arid, semiarid)	Bwh: Desert	Sudan, Doha, Central Australia	Sub tropical, erratic rainfall
	Bwk: Desert	China, Uzbekistan	Cold deserts
	BSh: Semi arid	Pakistan, Nigeria	Semi tropical
	BSk: Semi arid	USA (Denver)	Temperate steppe
C. Mild, humid (humid mesothermal climate)	Cfa: Subtropical wet climate	USA, Brazil	Hot and humid wet climate
	Cib: Temperate	Paris, Melbourne	Short cool summer
	Cfc: Sub polar	USA, Chile	Relatively cold winter
	Csa: Hot summer	Spain, Australia	Coldest month (−3 °C)
	Csb: Warm summer	Chile, South Africa	Short summer
	Csc: Cold summer	Hawaii	Coldest and driest months
	Cwa: Subtropical	Pakistan	Coldest and warmest months
	Cwb: Highland type	Latin America	Hot
	Cwc: Cold, subtropical	Bolivia, Peru	Coldest and warmest months
D. Continental	Dfa: Hot summer	New York	Highly variable weather
	Dfb: Warm summer	China	High rainfall in summer
	Dwa: Monsoon	Canada	Dry summer
	Dwb: Monsoon	Russia	Wet climate
	Dwc: Monsoon	Mongolia	Sub-Arctic climate
	Dwd: Monsoon	Russia	Dry and extreme cold weather

(continued)

Table 1.2 (continued)

Group	Type	Region	Characteristics
	Dsa: Mediterranean	Turkey	Coldest and driest months
	Dsb: Mediterranean	Turkey	Coldest and driest months
	Dsc: Mediterranean	Alaska	Coldest and driest months
	Dsd: Mediterranean	Siberia	Coldest and driest months
E. Polar	ET: Tundra type climate	Australia	Short summer
	EF: Polar climate	Antarctica	Eternal winter

transpiration over an area under sufficient moisture conditions. While classifying the climates, Thornthwaite used several factors as explained below.

Moisture: The precipitation alone is not a good indicator of moisture conditions, according to Thornthwaite, and, he has developed the concept of potential evaporation (PET) calculated from the temperature and day length. With the use of PE, in combination with evaporation, water budget methodology was used to develop moisture index. Moisture index was calculated from aridity and humidity indices.

$$\text{Aridity Index } (Ia) = (\text{D}/\text{PET}) \times 100$$

$$\text{Humidity Index } (Im) = (\text{S}/\text{PET}) \times 100$$

where

D it is the annual water deficit
S it is the annual water surplus
PET it is the annual potential evaporation.

Moisture index is defined as follows:

$$I_m = I_h 0.6x I_a = (100s{-}60d)/\text{n}$$

Using moisture index (I_m) climate of the world was grouped as moist and dry as per Table 1.3.

Thermal factor: Thornthwaite used "Thermal factor", and, termed it the "Index of Thermal Efficiency" (TE), on PE as e considered both TE and PE are of great importance in measuring the productivity of an environment. The climate classification as per TE is given in Table 1.4.

Table 1.3 The type of climate of the world and moisture index in Thornthwaite classification

Symbol	Climate type	Moisture index	Broad group
A	Per humid	100 and above	Moist climate
B4	Humid	80–100	Moist climate
B3	Humid	60–80	Moist climate
B2	Humid	40–60	Moist climate
B1	Humid	20–40	Moist climate
C2	Moist sub-humid	0–20	Moist climate
C1	Dry sub-humid	– 20 to 0	Dry climate
D	Semi-arid	– 40 to – 20	Dry climate
E	Arid	– 60 to – 40	Dry climate

Table 1.4 Climate type as per thermal index of Thornthwaite

Climate type	TE index × PE (cm)
A Megathermal	114 and above
B4 Mesothermal	99.7–114
B3 Mesothermal	60–80
B2 Mesothermal	40–60
B1 Mesothermal	57–71.2
C2 Microthermal	0–20
C1 Microthermal	– 20 to 0
D Tundra	– 40 to – 20
Frost	– 60 to – 40

Thornthwaite has considered moisture index above zero as moist climate (A, B4, B3, B2, B1 and C2) and moisture index below zero as dry climate (C1, D and E). He has provided subdivisions to express the extent of dry period under moist climate and extent of moist period under dry climate.

The subdivisions are given in Table 1.5.

Factors Which Influence World Climate Regions

Among the most important factors which influence world climate, the most important natural factors are the following:

1. Latitude
2. Altitude
3. Wind
4. Air mass
5. Distance from sea

Table 1.5 Sub divisions of moist and dry climate according to Thornthwaite classification

Moist climate description	Aridity index	Dry climate description	Humidity index
R Little or no water deficiency	0–10	D Little or no water surplus	0–16.7
S1 Moderate summer water deficiency	10–20	S1 Moderate summer water surplus	16.7–33.3
W1 Moderate winter water deficiency	10–20	W1 Moderate winter water surplus	16.7–33.3
S2 Large summer water deficiency	> 20	S2 Large summer water surplus	> 33.3
W2 Large winter water deficiency	> 20	W2 Large winter water surplus	> 33.3

6. Ocean currents
7. Direction of prevailing wind
8. Land topography
9. The El Nino phenomenon
10. Human effects.

Latitude (Distance from the Equator)

Due to curvature of earth, the temperature drops in areas far from the equator. This is because the rays of the Sun are dispersed over a larger area of land as we move away from the equator. This is very important, and, it is based on the concentration of sunlight and the area that it affects. In areas closer to the poles cause a higher albedo effect which reflects the solar energy for which it gives a cooling effect.

Altitude: Higher altitude usually have cooler temperatures and, generally, there is decrease of 1 centigrade Celsius for ever 100 m increase in altitude. Therefore, hilly at higher altitudes have cooler climate than in plains.

Wind and air mass: Heated ground causes air to rise which results in lower air pressure. When air rises it causes cooling and as it descends to the ground level it increases air pressure. These air masses absorb the climate of the air below it.

Distance from the sea (continentality): Oceans heat and cool down much slowly than the inland areas. This is the reason why coastal locations tend to be cooler in summer and warmer I winter than inland areas at the same altitude and latitude.

Ocean currents: The rotation of the Earth affects the oceans setting up ocean currents which flow within the ocean basins. Ocean currents are driven by surface winds, Earth's rotation and differences in salinity. Under the influences of the trade winds, surface waters near the equator flow from east to west. Warm water pools along the west coast of continents, which sets up a temperature gradient across the ocean surface. Disruption of this temperature gradient creates the event known as El Nino. El Nino refers to the irregular warming of the surface water in the Pacific.

Human effect: Apart from natural factors described above, human effect also plays a vital role in climate pattern. With population rise, there is increased deforestation leading to enhanced carbon di oxide concentration in the atmosphere. Emission of Green House Gases (GHGS) such as, methane, nitrous oxide, increases also cause change in climate as clearly described in the book of Nair (2019) titled "Combating Global Warming: The Role of Crop Wild Relatives for Food Security".

References

Koppen, W. 1936. Das geographische System der klimate. In: Handbuch der Klimatologie. Vol 1, Part C (W. Koppen and R. Geiger, Eds.) Berlin, Gebruder Borntrager

Nair,K.P.P. 2019. Combating Global Warming - The Role of Crop Wild Relatives For Food Security. Springer Nature Switzerland AG 2019

Thornthwaite, C.W. 1948. An approach towards a rational classification of climate. Geographical Review. 38, 55–94

Chapter 2
Climate and Crop Production

Climatic Factors and Crop Production

Crop growth depends on both internal and external factors. The internal factors are mostly genetic or hereditary, in nature, which include high yielding capacity, resistance to lodging, tolerance to abiotic conditions like drought, flood and salinity, and also, tolerance to biotic conditions like both insect and pest attack. The external factors include climate, edaphic, biotic, physiographic and socio-economic factors. Climate influences the yield potential of a crop. Nearly 50% of the crop yield is attributed to climatic factors. The climatic factors are mostly abiotic which affect plant growth and development. Such factors which affect plant growth, development and crop productivity are, water availability, temperature, solar radiation, atmospheric humidity, wind velocity and atmospheric gases. However, solar radiation, temperature and rainfall are the most important of these. These factors influence the growth, development and yield of the crop. They are discussed below.

Solar Radiation

It is the source of all hear energy (about 99.97% of total Earth energy), which is required for all of the physical processes taking place in the Earth-atmosphere system. Solar radiation provides light, which regulates the seed germination, leaf expansion, growth of the shoot, flowering, fruiting and other physiological functions in the plant. Effect of solar radiation on plants can be categorized into four, as follows:

1. Thermal Effect: The bulk of the solar radiation absorbed by plants is converted to heat energy, which is used for transpiration. Temperature obtained from solar radiation will be discussed later in this chapter under the head "Temperature".

K. P. Nair, *Climate Resilient Sustainable Agriculture*,
https://doi.org/10.1007/978-3-031-84166-8_2

2. Photosynthetic Effect: About 28% of solar energy is utilized for photosynthesis in pants for carbohydrates production. Light quality, intensity, and day length are important factors regulating plant growth and plant development.
3. Photoperiodic Effect: Solar radiation regulates and controls the flowering and vernalization in plants according to length of the light period.
4. Other Effects: Solar radiation influences assimilation of plant nutrients and dry matter distribution.

Solar energy is an electromagnetic phenomenon which bears the characteristics to a body at 6000° and is the result of thermonuclear fusion of many hydrogen atoms. More than 99% of the solar energy is contained within the wave length region of 150–4000 nm (0.15–4 μm) which is called short wave radiation. The distribution is 9% below wave length of 400 nm (ultraviolet), 45% within the visible range of 400–700 nm (blue, green, yellow, red) and 50% is in the infra-red range of more than 700 nm (far-red and infrared). The spectral distribution varies with altitude. The solar light within the range of 400–700 nm is effective for photosynthesis. The solar radiation between 315 and 400 nm produces formative effects and 700–920 nm has important role in photoperiodism, germination of seeds, flowering and coloration of fruits. Wave length below 315 nm has detrimental effects on plants.

The solar radiation on the top of the atmosphere depends on latitude, season and day/time/h. Solar radiation falling on the Earth is also reflected which is called albedo. Albedo is measured as the ratio of incident solar radiation reflected by a body and expressed as a percentage. The emissivity varies with the nature of the surface. The values given in Table 2.1 of E are given by Brooks and Geiger.

During photosynthesis, carbon dioxide is reduced to carbohydrates in the presence of catalytic enzymes, such as, phosphenol pyruvate carboxylase and ribose biphosphate carboxylase whose activity are is regulated by light intensity. Different pigments required for photosynthesis are also produced in the presence of light. Since solar radiation influences crop growth, directly, its reduction will appreciably reduce agricultural productivity. The accompanied increase I minimum temperatures increases respiration requirement of the crops and thus reduces the net growth and productivity (Aggarwal 2003). Pathak et al. (2003) reported the negative trends in solar radiation and an increase in minimum temperature, resulting in declining trends of potential yields of rice and wheat in the Indian Indo Gangetic Plains.

Table 2.1 Values of E and albedo

Surface	Infrared emissivity	Albedo (less than 4 μm)
Water surface	92–96	6–9
Snow	82–99	75–95
Soil dry, dark	90	5–15
Wet grass	–	33–37
Dry grass	–	15–28
Earth	–	34–42

Solar radiation interception is not uniform in al plant parts. Distribution of light within the plant community is different from its distribution on a single plant or leaf. Leaf orientation and leaf area index (LAI) are important factors for interception of light and radiation. Higher the LAI, required to intercept the same fraction of light, higher the crop efficiency in utilizing solar energy. Radiation profile also changes with plant height day time hours. Generally, maximum interception is in noon, and, minimum in morning and sun set. On average, 75% of incident radiation is absorbed by crop canopy, 15% reflected and 0% transmitted from the crop field. Efficient solar radiation utilization takes place when 13% of leaves are oriented at 0–30°, middle 37% at 30–60° and upper 50% at 60–90° orientation. Actual radiation utilization by most crop plants is less than 1% and efficiency varies from crop to crop. Efficiency Utilization efficiency in potato is 0.5%, in wheat 0.90%, maize 1.05%, and, 1.43% in sugarcane. Normally, C_4 like sugarcane and maize are more efficient in utilization of solar energy compared to C_3 plants like wheat and rice. The ratio of photosynthetically active radiation (PAR) is close to 0.50 in tropical and temperate regions (Montheith 1972) and represent a weighted mean between functions of direct radiation and different sky radiation.

Direct linear relationship exists between light intensity and photosynthetic rate. At light intensity of 500 ft. candle net assimilation would be zero and a minimum of 500–1000 ft. candle is required for effective photosynthesis. Extremely high light intensity will have a deleterious effect on photosynthesis, which is called solarization. Different plants require different light hours to flower. Long day plants require more than 14 h, short day plants require less than 10 h, and, day neutral and indeterminate plants require 12–14 h, and flowering initiation occurs at different periods.

Water

Water is the elixir of life. Plants require huge quantities of water which is required to photosynthesize, respire, absorb, translocate and utilize plant nutrients, divide cells, and for many other vital processes. It also affects evapotranspiration which is a combined process of evaporation of water from the surface of the Earth and the surface of the plant and transpiration of water through the plant stems and tissues. Although a plant hardly utilizes 1% of the water absorbed, for its metabolic processes, evapotranspiration contributing to as much as 99% of water loss is a "necessary evil". Water affects the performance of plant not only directly, but also, indirectly, influencing the bioavailability of plant nutrients absorbed, the timing of cultural operations in a field. Water is a critical input impacting crop yield. Crop yield is impacted both by excess and/or scarcity of water in the soil.

Precipitation includes rainfall, hail, fog, and dew, but, rainfall is the principal source of water for the plant, affecting plant growth, development and yield. Hence, rainfall distribution is decisive of the nature of cropping in a given region. However, crop yield is not directly proportional to the total quantum of rainfall received, but, how well it is distributed. A skewed distribution will do much harm to the crop. For instance, excess rainfall immediately after planting/sowing seeds or in early stages of plant growth could be detrimental in not letting the plant grow properly, similarly, excess rainfall during harvest will lead to much crop loss. The rain that falls on the ground may be effective to the maximum. To be reasonably safe from evaporation, the raindrops must penetrate to a depth of at least 10–12 cm of the soil (Staple 1964). Annual effective rainfall is about one-third of the total (Bagnold 1954). Proper analysis of the rainfall pattern is a pre-requisite for efficient crop planning. Rainfall distribution is more important than total rainfall received. For instance, 300 mm of rainfall on a single day, followed by 2–3 weeks of no rainfall, will cause a drought-like situation, whereas the same quantity of evenly distributed rainfall during the critical period of crop growth will be like a boon to enhance crop yield. Dependability of rainfall is estimated by 75% of probability rainfall and by the coefficient of variation. Length of Growing Season (LGS) also is rain-dependent. It is the period between seed to seed, when the soil profile moisture is reduced to a ratio of AE/PE to 0.5, where AE and PE are actual and potential evaporation rates, respectively. Length of growing period during the year is the number of days when the precipitation exceeds 0.5 of PE plus a period required to evapo-transpire an assumed 100 mm from the soil profile storage.

Hydrological Cycle

Water cycle or hydrological cycle is the normal water recycling system on planet Earth. Sea water, lake water, and water from other bodies evaporate under the impact of solar radiation. Water also evaporates from plant leaves through the mechanism of transpiration. The water vapor cools in the atmosphere, condense as clouds and return to the land and the sea, as precipitation (rainfall). Precipitation falls on the Earth as surface water, which flows off as streams of water that result in lakes and rivers. A part of the water that precipitates infiltrates into the ground and moves downwards through the incisions, forming aquifers. Finally, a part of the surface and underground water moves into the sea and remains as liquid gas and ice.

Available Soil Moisture

Rain water or irrigation water is stored in the soil which is require for plants to grow. All the water that enters the soil is not fully available for the plants growing in the soil. Soil water is found as, saturation, field capacity, wilting point and vapor. When rain falls on the ground, some of it is immediately lost through evaporation, and a portion as "run off". Remaining quantity enters the soil. Following rainfall or irrigation, field soil becomes saturated as all the soil pores (air spaces) are filled with water. Following saturation, a portion of water drains into the underground under gravitational pull, which is called gravitational water. Field capacity is the amount of water which remains after loss of gravitational water and is regarded as the upper limit for crop use. It is expressed in terms of energy status as one-third atmospheric tension or 0.1–0.33 bar. Once soil water is gradually used, by crop plants, without further replenishment, of water, either through rain or irrigation, the plant wilts as it is unable to meet the evaporative demand. This state is called the wilting point, which is either temporary or permanent. At permanent wilting point, the energy status of soil moisture reaches—15 bar. At this point, plant wilts and survives as it is able to draw small amount of water from the soil during cool periods. Further drying of soil, water is held in soil very tightly and the plant roots are unable to absorb water and starts to dry. The soil moisture at this stage reaches—60 bar and water is retained in the form of vapor. The limits of soil water between the field capacity and permanent wilting point/coefficient (FC-PWC/P) is called the relatively available moisture (RAM) for the pant/crop. Moisture extraction from different soil depths by pant roots is non uniform. Maximum water is extracted from the upper zone of soil. Approximately, 40% soil water is extracted from the upper zone, followed by 30%, 20% and 10% respectively, from the second, third and fourth zone.

Temperature

On a global scale, temperature influences the distribution of both animals and plants. It influences, evapotranspiration, photosynthesis and soil warming. Soil temperature is more important that atmospheric temperature, with regard to seed germination, nutrient absorption, root growth etc. Air temperature impacts all growth processes in plants, such as, respiration, transpiration etc. For all vital activities such as respiration, transpiration, photosynthesis and nutrient absorption and translocation within the plant, a plant requires an optimum temperature, which is mostly, between 15 and 45 °C. High temperature adversely affects respiration and photosynthesis. Water absorption is inhibited when soil temperature is low, as water gets more viscous at cool temperatures.

Following are the major impacts of temperature on plant growth and development:

1. The decomposition of soil organic matter: When microbial activity is higher it presupposes that the soil temperature conditions are ideal.
2. Plant growth and development: At ideal soil temperature, activity of auxins, gibberellins (responsible for plant stem elongation), and cytokinins (responsible for plant cell division), and also, growth regulators like ascorbic acid are increased.
3. Respiration: Under increased soil temperature respiration is enhanced. High night soil temperature leads to exhaustion of plant assimilates. Poor net accumulation results in poor crop yields.
4. Biochemical reaction: The enzymes which regulate vital plant activities are regulated by soil temperature. Usually, enzymatic reaction is doubled with enhanced temperature of soil by 10 °C (Mader 1993). This is called Q_{10}.

5. Pest and disease attack: High soil temperature and humidity enhance incidence of both.
6. Photosynthesis: Soil and atmospheric temperature regulate protoplasts formation. At low temperature leaves turn yellow, due to degradation of chlorophyll. Recovery rate of photosynthesis is uniform in all leaves after the low temperature stress is eliminated.

Temperature influences seed germination and different seeds have different temperature requirements as shown in Table 2.2.

Temperature of soil is slightly more than the atmospheric temperature due to high specific heat of the soil. In India, air temperature is maximum in the summer month of May, while it is lowest in the winter months of December–January. The difference in maximum and minimum temperatures is minimum in the month of September due to higher moisture content in soil profile, as compared to other months, when the soil is dry. This indicates upward movement of heat from the soil to atmosphere.

Plants grow less satisfactorily under constant temperature than when they are exposed to day-night temperature differential. This response is called "thermoperiodism", which varies with the stage of the particular crop growing in the field, and, from one crop to another. Low night temperature increases the sugar content of the plant under consideration and translocation of sugars is favored in a temperature regime of 25 °C.

The temperature requirement of some of the common crop species is given in Table 2.3.

Higher temperatures have more deleterious effects on net photosynthesis than low temperatures which lead to decreased production of photosynthates above the critical temperatures. Temperature often controls flowering as affected by photoperiods. Information on accumulated day degree requirement for different crop cultivars allows sowing days to be planned well ahead of the sowing season (Reddy 1995).

Table 2.2 Effect of temperature on percentage germination of seeds of different crops

Crop	Soil temperature in Celsius and germination percentage								
	5	10	15	20	25	30	35	40	45
Rice	0	0	92	100	100	96	85	60	0
Wheat	40	97	97	100	100	97	85	18	0
Maize	0	0	93	100	100	96	85	60	0
Sorghum	0	0	55	95	98	98	98	83	0
Cotton	0	0	18	93	96	98	98	83	0
Gram	35	98	100	100	100	100	100	55	0
Pea	38	98	100	98	98	63	0	0	0
Bottle gourd	0	0	0	44	97	97	97	53	0
Okra	0	0	0	73	100	98	100	85	0

Source Singh, N. I. and Dhaliwal, G. S. 1972 Plant and Soil Vol 37, 441–444

Table 2.3 Temperature requirement of some selected crops

Crop	Minimum temperature	Optimum temperature	Maximum temperature
Rice	10	32	36–38
Wheat	4.5	20	30–32
Maize	8–10	20	40–43
Sorghum	12–13	25	40
Tobacco	12–14	29	35

Note Temperatures are in Celsius

Atmospheric Humidity

Relative Humidity (RH) has direct effect on evapotranspiration and indirect effect on growth and development, photosynthesis, pollination, and disease and pest damage in the field, which ultimately affects the crop yield. In dry regions, the RH remains low, as a result of which there will be water deficit in plants leading to reduced leaf water potential.

High RH increases the incidence of pests and diseases, reducing the efficiency of plants to intercept solar radiation. Low RH increases evapotranspiration leading transpiration and leaf wilting. The amount of water in the air rages from 0.01% by volume at the frigid poles to 5% in the humid tropics. The air becomes moist in high RH, while it is dry in low RH. The RH affects the opening and closing of the leaf stomata which regulates the water from the plant through transpiration and photosynthesis. For plant propagation, humid conditions are most favorable. High RH favors leaf enlargement, while moderate RH favors pollination and seed setting. The water requirement of the crop is also influenced by RH. Most of the crops perform between RH of 40–60%, while a few crops perform well at RH 80%.

Wind Velocity

The motion of air is called wind. Winds blowing over a warm current are usually laden with moisture, while those blowing over a cold current have a cooling effect on the coast leading to the formation of fog. Wind turbulence in the atmosphere favors the exchange of gases. The carbon dioxide and oxygen contents of air is very important from the point of view of photosynthesis and respiration. Wind moving at a great speed enhances the rate of evapotranspiration and reduces the water balance in the leaves due to the reduction of relative turgidity. When transpiration rate exceeds water absorption rate, by roots, the leaf stomata may be closed and it will restrict the diffusion of carbon dioxide. This will cause reduced photosynthesis, growth and yield. Wind over the crop surface can alter the onset of drought during dry periods due to water vapor loss through transpiration. This leads to stomatal closure and reduced rate of gaseous exchange leading to photosynthesis reduction and reduction in crop growth. Moisture stress in the leaves causes anatomical and morphological modifications in plants due to moisture reduction. Winds can alter the hormone balance as well within the plant system. Hot and dry winds cause plant height reduction in cell expansion. High velocity wind causes plants to lodge. Mechanical damage may also be caused to leaf tips by strong winds.

Wind classification is done taking into account wind speed at km/h. Wind blowing at less than 1.5 km/h is classified as "calm wind", while that which blows at 100–120 km/h is classified as "storm" and wind blowing at a speed exceeding 120 km/h is termed "hurricane". Gas exchange takes place when winds bow gently at a speed of 4–6 km/h and winds blowing at higher speeds can cause mechanical damage to both plant leaves and plant stem. Plant growth is inhibited when wind speed exceeds 10 km/h (Went 1957). Favorable wind speed is essential for dispersal of plant pollen for fertilization and see setting in pants. Excessively speedy winds, such as storm and hurricane, can also cause soil erosion.

Monsoon is a flow pattern of general circulation in which there is clearly a dominant wind in one direction in every part of a region. In India, there are two monsoons, South West (starting from June–July and closing by September/October) and North East monsoon during winter months. The onset and closure dates differ in India from region to region.

Atmospheric Gases

Most of the dry air is in the troposphere (troposphere is the lowest and densest part of the earth's atmosphere) in which most weather changes occur and temperature generally decreases rapidly with altitude and which extends from the earth's surface to the bottom of the stratosphere at about 7 miles (11 km) high. Stratosphere is an upper portion of the atmosphere above the troposphere where temperature changes little and clouds rarely form. Most of the dry air in the troposphere consists of 78.09% nitrogen (elemental), 20.95% oxygen with 0.92% argon, 0.03% carbon dioxide, as well as traces of hydrogen and other gases.

Carbon dioxide is a raw material for photosynthesis in plants, while it is also a byproduct of respiration. Oxygen is a byproduct of photosynthesis, but an input for respiration. Atmosphere can be polluted by suspended particles of dust and presence of carbon monoxide (CO), carbon dioxide (CO_2) m sulfur dioxide SO_2 and sulfur trioxide SO_3, nitrogen oxides, methane CH_4, chlorofluorocarbon (CFC), and some suspended particles. Polluted air can adversely affect the rate of photosynthesis and respiration, which ultimately affect the plant growth. Ozone depletion in the atmosphere will lead to increase in ultra violet radiation (Nair 2019), with very drastic adverse impact on the Earth's environment, especially due to anthropogenic heating (Nair 2019). Acid rain is often occurs due to high concentration of sulfur dioxide and sulfate in the atmosphere.

Most plants growing in enhanced CO_2 concentration usually enhances net photosynthesis. The higher photosynthetic rates are then manifested in higher leaf area, dry matter production, and yield of many crops (Cure 1985). CO_2 enrichment reduces transpiration per unit area also. Higher CO_2 often improves water use efficiency. Temperate crops are more efficient than tropical ones in CO_2 utilization. Crops, such as, rice, wheat, cotton using C_1 pathway of photosynthesis are found to be less efficient in photosynthesis than C_2 pathway of photosynthesis than C_4 pathway crops, such as, maze, sugarcane, sorghum and millets.

Nitrogen is one of the major plant nutrients, and, elemental nitrogen, present in abundance in the atmosphere (78–80%), cannot be directly used by cereals like rice, wheat, maize etc. Atmospheric nitrogen is fixed in the soil by lightning and first rainfall. However, most elemental nitrogen of the atmosphere is fixed by the bacteria *Rhizobium meliloti*, for example, found on roots in leguminous plants, for instance, beans. However, non-symbiotic anaerobic bacteria, such as, *Clostridium*, *Azobacter*, and, *Azospirillum* etc., are free-living with N fixing ability. Both carbon cycle and nitrogen cycle are important biological cycles which function in the soil.

Weather and Crops

In addition to the soil factors, atmospheric variables, such as, solar radiation, air temperature, humidity, and precipitation impact crop performance. Field operations are also influenced by weather factors. Weather aberrations, such as, intermittent dry spell and early withdrawal of monsoon, excess or deficit of rainfall, temperature fluctuation, overcast skies etc., can affect growth and development of crops.

Dry soil will result in poor plant establishment after sowing, while wet soil will result in nitrogen loss, through leaching and denitrification. This can also lead to development of seed, root, and crown diseases. The impacts on crop yield due to high temperature extremes have been modelled or empirically estimated on a global scale during the past few years (Derying 2014). Tropical cyclone may cause crops loss due to blowing tides, insufficient oxygen caused by overhead flooding, flash floods, wind injury to plant organs and water stress induced by enforced respiration, all of which occur simultaneously (Masutomi 2012).

References

Aggarwal, P.K. 2003. Impact of climate change on Indian Agriculture. J. Plant Biology 30, 189–198

Bagnold, R.A. 1954. The physiological aspects of deserts. In: Biology of deserts (Coundsley-Thompson, Ed.). Institute of Biology, London

Cure, J.D. 1985. Carbon dioxide doubling response: A crop survey: In: Direct Effects of Increasing Carbon Dioxide in Vegetation: B.R. Strain and J.D. Cure (Eds). DOE/ER – 0238, US Dept. of Energy, Washington D.C. 99–116

Derying, D. 2014. Global crop yield response to extreme heat stress under multiple change futures: Published 19 March, 2014. IOP Publishing Ltd Environmental Research Letters, Volume 9, November 3

Mader, S.S. 1993. Biology Part I: The Cell. 4th ed. Dubuque, IA, USA: W.C. Brown Communications Inc. 152pp

Masutomi, Y. 2012. Estimation of the damage area due to tropical cyclones using fragility curves for paddy rice in Japan. Environ. Res. Lett. 7 (2012) 014020 (9pp) https://doi.org/10.1088/1748-9326/7/014020

Montheith, J.L. 1972. Solar radiation and productivity of tropical ecosystem. Journal of Applied Ecology. 9: 747–766

Nair, K.P.P. 2019. Combating Global warming: The Role of Crop Wild Relatives for Food Security. Springer Nature Switzerland AG 2019

Pathak, H., Ladha, Aggarwal, P.K., Peng, S., Das, S., Singh, Y., Singh, B., Kamra, S.K., Mishra, B., Shastri, A.S.R.A.S., Aggarwal, H.P., Das. D.K., and Gupta, R.K. 2003. Trends of climatic potential and on-farm yields of rice and wheat in the Indo-Gangetic Plains. Field Crops Res. 80, 223–234

Reddy, S.R. 1995. Principles of Agronomy, Kalyani Publishers, Ludhiana, India

Staple, W.J. 1964. Dryland Agriculture and Water Conservation. In: Research on water. ASA Special publication. Soil Science Society of America Proceedings. Madison, Wisconsin

Went, F.W. 1957. Experimental control of plant growth. Chronica Botanica 17, Ronald Press Co. New York, USA

Chapter 3
Impacts of Climate Change on Agriculture

In a particular region for a crop to be grown, one needs, good soil, good climate and sufficient water. Climate affects temperature trends, seasons, length of the growing season, amount of sunlight and seasonality of sunlight etc. A change in climate may cause recurrent drought, floods and weather aberrations that would affect agricultural production, ultimately human sustenance. The relationship between climate and agricultural production is very complex as it is dependent on climate which also plays its role in the emission of greenhouse gases. Nair (2019) has elaborately discussed this aspect in his book "Combating Global Warming: The Role of Crop Wild Relatives for Food Security" published by Springer Nature.

Climate change is experienced by all and everyone feels its effects. The farmers are already experiencing the adverse effects of climate change. There are many deviations in weather patterns. Temperature increase is 1.1 °C above pre-industrial level due to emissions of green house gases (GHGs). When these gases are emitted, into the atmosphere, they trap heat close to the earth's surface due to the green house effect. This results in global warming. The phenomenon is clearly explained (Nair 2019). More intense rainfall, longer dry spells, and other natural disasters are the result of climate change. Extreme climate change affects both crop production and livestock farming.

Which Are the Factors that Contribute to Climate Change?

Climate change is caused by:

1. Geoecological Factors
2. Anthropogenic (Human Induced) Factors.

K. P. Nair, *Climate Resilient Sustainable Agriculture*,
https://doi.org/10.1007/978-3-031-84166-8_3

Geoecological Factors: There are several, such as, earthquake, tsunami, global warming/cooling, global dimming, and incursion of sea water into the land mass, which cause climate change. Some of the important ones are discussed below:

a. Continental drift: Continental drift has an impact on climate change because it changes the physical features of the landmass, their position, and the position of water bodies. The separation of the landmasses changed the flow of ocean currents and winds, which affected the climate.
b. Volcanoes: When there is an eruption of a volcano, it throws out sulfur dioxide (SO_2), water vapor, dust and ash into the atmosphere. Volcanic activity lasts only for a few days, but, the quantum of gases and ash it releases can vastly influence climate patterns for years. Large quantity of sulfur dioxide can reach the stratosphere (stratosphere is the part of the earth's atmosphere) which extends from the top of the troposphere (the lowest densest part of the earth's atmosphere) in which most weather changes occur and temperature generally decreases rapidly with altitude and which extends from the earth's surface to the bottom of the stratosphere at about 7 miles (11 km) from a major eruption. These gases and dust particles partially block the incoming rays of the sun, leading to cooling. Sulfur dioxide combines with water to form tiny droplets of sulfuric acid (H_2SO_4). These droplets ae so tiny that many of them can stay in the sky for years on end, and, are efficient reflectors of sunlight and screen the ground from some solar energy.
c. The earth's tilt: Each year the earth makes ne round the sun. It is tilted at an angle of 23.5° to the perpendicular plane of its orbital path. For one-half of the year, when it is the summer period, the northern hemisphere tilts towards the Sun. In the other half, when it is winter, the earth is tilted away from the sun. Such tilt of the earth causes seasons like summer and winter. More tit causes warmer summers and colder winters, while less of it causes cooler summers and milder winters. The gradual change in the direction of the earth's axis is responsible for changes in the climate.
d. Ocean currents: The oceans cover about 71% of the earth and absorb about twice as much as the Sun's radiation as the atmosphere or the land surface. Ocean currents move vast amounts of heat across the planet which is approximately the same as the atmosphere. As the oceans are surrounded by land masses, heat is transported through channels. Winds push horizontally, against the sea surface and drive ocean current patterns. The El Nino event in the Pacific Ocean can affect climatic conditions of the world. Ocean currents may change or slow down. Much of the heat which escapes from the oceans is in the form of water vapor. This is also the most abundant greenhouse gas on earth and plays a crucial role in climate change. However, water vapor also contributes to cloud formation.
e. Earthquakes: Both human activity and natural factors cause earthquakes. Sudden release of stored energy on the earth's crust create seismic waves which results in earthquakes, characterized by sudden shaking or displacement of the ground. An earthquake caused by tectonic plates points a strain on the ground which becomes so great that rocks give way to breaking and sliding along faulty planes.

Sometimes earthquakes cause tsunamis that may case property loss and human lives. Smaller earthquakes can be caused by volcanic activity, landslides, mine blasts, and nuclear experiments.

f. According to latest research, methane release from earthquake contributes to climate change. Methane is held in pockets generally located at sub-sea level and can be released by the shifting tectonic plates. Methane can be released from sub-sea pockets due to earthquakes, as per the investigations of German and Swiss scientists. As per many investigations, there is a correlation between earthquakes and increase in global temperature.
g. Tsunami: With the rapid displacement of water from the seas, a series of waves are crated which are called tsunami. Earthquakes, volcanic eruption, mass movement of water from ocean, explosion and landslides create tsunami.
h. Global warming: Over the last century, there has been a temperature increase of air near earth's surface by about 1 °C. Since the earth is heated by this many disasters such as hurricanes, droughts and floods occur.
i. Global cooling: It is the result of extensive glaciation and the cooling effects of aerosols or orbital forcing. Paul Ehrlich (1968) mentioned global warming from greenhouse gases as a counterforce to the cooling effect of aerosols. However, the possibility of this occurrence is generally very low. There is now general agreement that aerosol effects were the dominant cause for the mid twentieth century cooling. Due to fossil fuel combustion, tiny dust particles, called aerosol, re increased in the atmosphere. Aerosols directly increase the planetary albedo reducing solar radiation from reaching the surface and causes cooling.
j. Global dimming: Global reduction in irradiance at the surface of the earth is called global dimming, which is caused, by increase in particulates such as sulfate aerosols in the atmosphere.

Human Factors

Among the human factors, the anthropogenic warming caused by human factors (Nair 2024), and the industrial revolution and release of green house gases (GHGs) are the major causes for climate change.

Industrial revolution: Industrial revolution has shown to be important for economic "development". But, the resultant climate change is cause for concern. The green house gases originate from water vapor, carbon dioxide, methane, nitrous oxide and chlorofluorocarbons (CFCs).

Along with industrialization, the following activities have also been responsible for climate change:

1. Power generation: Power generation by burning fossil fuel, such as coal and natural gas, produce GHGs, such as, nitrous oxide and carbon dioxide which blanket the earth and trap radiation from the sun leading to increase in atmospheric temperature.

2. Manufacture of goods: Production of cement, ion, steel, electronics, plastics and other goods including mining and industrial manufacturing releases huge amounts of GHGs into the atmosphere leading to atmospheric temperature increase.
3. Transportation: Cars, truck, ships, and planes, which run on fossil fuels, contribute appreciably GHGs leading to enhancement of atmospheric temperature.
4. Powering buildings: Electricity produced form coal, oil or natural gas is used for heating buildings in Europe and also for cooling in tropical countries, which also leads to emission of GHGs and lead to increase in atmospheric temperature.

Intensive or Industrial Agriculture (read "Green Revolution"): Possibly, the biggest contributor of GHGs, in particular, due to emission of nitrous oxide through unbridled urea fertilizer use to prop up the green revolution, as has happened in the states of Punjab (the "cradle" of Indian green revolution) and Haryana, in India, due to microbial decomposition (Nair 2019).

Agriculture, forestry, and other land use (AFOLU) sectors are the second most important source of emissions contributing around 24% of the emissions, whereas electricity production, industry, transport, buildings and other sectors contribute 25%, 21%, 14%, 6.4% and 9.6% respectively. The GHGs of relevance here are methane, carbon dioxide and nitrous oxide.

Soil emissions: 60% of the total emissions through the agricultural sector is through nitrous oxide emissions. During the green revolution phase this source contributed about 30–35% of global (anthropogenic) warming (Nair 2024). Unbridled use of nitrogenous fertilizers, such as urea, to prop up the green revolution is the principal source of this emission because of microbial decomposition of the synthetic fertilizer.

Enteric fermentation: During digestion of the feed, microbes in the animal's digestive system ferment the feed. This process of enteric fermentation produces methane as a by-product which can be emitted by exhaling and belching of the animal. Large ruminants like cows and bullocks have higher methane emissions than pigs and poultry, because of their unique digestive systems.

Manure management: Methane is also produced by the anaerobic decomposition of manure. When manure is handled as a solid or deposited naturally on grassland, it decomposes aerobically and creates methane emissions.

CO_2 from fossil fuel consumption: Combustion of fossil fuel like gasoline and diesel, used to fuel farm implements, including tractors, combines etc., produce CO_2. It also includes emissions related to synthetic fertilizer production, pesticides and herbicides, which are primarily derived from fossil fuels.

Rice cultivation: Rice is a water-guzzling plant, and, mostly it's grown in water-logged (submerged) conditions. Flooding causes decomposition of manure, and other organic manure (like farm yard manure, green manure, vermicompost etc.), and decomposition of these materials result in methane emission. This is released into the atmosphere.

Fertilization: A relatively new GHG nitrous oxide, which can occur naturally, but, its content has markedly increased as a result of increasing use of synthetic fertilizers. The direct emissions related to agriculture worldwide is estimated to be 3.5% of all GHG emissions.

Deforestation: According to FAO, deforestation, worldwide, mostly related to agriculture, is estimated to be responsible for 17.4% of total GHG emissions. Deforestation causes emissions when trees with stored carbon are cut. Since forests absorb CO_2, destroying them also limits the ability of nature to keep emissions under check. Deforestation also leads to loss of biodiversity and the degradation of the environment (Nair 2023). Without compensatory afforestation, deforestation leads to loss of habitats and biodiversity reduction (Nair 2023).

Water storage in dams: Reservoirs storing water for irrigation will influence the macro-climate of the adjoining area, in addition to waterlogging, and salinity in these areas.

Greenhouse effect: The energy received from sun as ultraviolet rays drives earth's climate. Most of the energy is radiated back to the space, but, some is trapped in the ground, the ocean and the atmosphere. Nitrogen, oxygen and argon are the major atmospheric gases which directly affect climate. GHGs act like the glass in a greenhouse and block the heat which warms up the atmosphere. This is known as the greenhouse effect. Earth's atmosphere has always acted like a greenhouse to capture sun's heat and influence life forms.

CO_2: Different GHGs have different effects depending on their concentration in the atmosphere, ability to alter radiation balance and time of availability. To make comparisons of the global warming effects of different gases, the IPCC (International Panel of Climate Control) uses global warming potential (GWP) as an index to "represent the combined effect of the differing times these substances remain in the atmosphere and their effectiveness in causing radiative forcing". The larger the GWP, the higher is the warming effect of a given gas, compared to CO_2, over a time period, usually in terms of 100 years. For instance, one ton of methane is equivalent to 25 tons of CO_2, while one ton of nitrous oxide is equivalent to 298 tons of CO_2 (IPCC 2007).

A molecule of CO_2 has the smallest effect on the climate, but, it is by far the most important anthropogenic GHG, hence, has the biggest overall effect. The GWP of CO_2 is 1 regardless of the time period and t is used as the reference which allows comparison with other gases. CO_2 accounts for about three-fourths of total emissions and its overall contribution to radiative forcing continues to rise. It is difficult to provide a single lifetime of CO_2 because the gas not destroyed over time, rather, it moves through the earth system. Most of it's excess has been incorporated in the oceans over decades while the balance will remain in the atmosphere for thousands of years to come.

Methane is a more potent GHG than CO_2, but, sunlight converts it to CO_2 after about 12 years. Averaged out, the GWP of methane over 100 years is 28.

Other gases are more powerful sources of warming and last much longer in the atmosphere. Nitrous oxide lasts 121 years and has a GWP of over 100 years of 265, while some types of fluorinated gases may last thousands of years with a GWP over 100 years of more than 6500.

Since all systems in the global climate system are interconnected, adding heat energy causes the global climate to change, as a whole. Much of the world is covered with water (oceans) which heats up and more water evaporates into clouds. Where storms like hurricane and tornadoes, typhoons are forming, the result is more energy-intensive storms. A warmer atmosphere makes glaciers and mountain snow packs, the Polar ice cap, and the great ice shield jutting out of Antarctica, melt raising sea levels.

Which Are the Main Green House Gases (GHGs)?

Anthropogenic factors (human factors) are the main sources for the emission of GHGs. The quantum of CO_2 released by each country depends on the type and share of its main economic activities. The latest estimate shows a total global emission of 46 billion of GHGs expressed in terms of CO_2 equivalents. This is an increase of 35% over previous decade. Electricity and heat generation (to keep human dwelling in cod countries) are the biggest sources of GHGs. Agriculture, forestry, and other land uses (AFOLU) represent the second most important source of emissions, amounting to 24%. Industry accounts for 21%, others which these include transport by motor vehicles, trains, ships and air craft, all powered by fossil fuels, account for 14%.

Impact on climate change: Climate change affects plants and animals by changing their natural equilibrium directly or indirectly. Loss of forest cover causes soil erosion directly and adversely impacts animal habitats through their physical displacement.

What Are the Impacts of Climate Change at Global Level?

The following impacts of climate change are felt on various human activities (anthropogenic):

Water: Water availability will be severely affected by both drought and flooding.

Food: Increased temperature and frequent droughts and floods due to climate change will reduce crop yield, consequently food availability to mankind. High CO_2 levels will enhance photosynthesis leading to greater production of plants, in particular, in C_3 plants. The C_3 plants ae, rice, pulses, oilseeds and vegetables. Increased CO_2 will also reduce produce quality, in terms of protein, nitrogen and other mineral contents, in most plant species. Incidentally, climate change will also impact weed growth. Both historical and future estimated impacts of climate change on important cereals like rice is – 35%, – 20% wheat, – 50% sorghum, – 60% maize and – 13% for barley. These reductions are also dependent on where these crops are grown.

Industry, settlement and society: Industries located in coastal areas and river flood plains will be affected more by weather extremes. Due to more intense weather extremes, economic and social costs of life will escalate.

Health: Heat waves, floods, tornados, hurricane, cyclone, storms, fires and droughts caused by climate change may affect the health of mankind, causing increased deaths and diseases. The problems of malnutrition, diarrhoea and malaria in some areas/ regions will escalate.

Warming and sea level: It is projected that throughout the twenty-first century, India will experience warming above global level. It is also expected that India will experience more seasonal variation in temperature with more warming in the winters than in the summers. In recent years, extreme heat waves across India extended together with warmer night temperatures and hotter days, and this trend is expected to continue.

Natural calamities: Due to rainfall variability, changes in the frequency and severity of droughts and floods could pose challenges for farmers and ranchers and will, ultimately, threaten global food security. Meanwhile, rise in sea level and waring of sea water are likely to cause habitat changes of many fi species, disrupting ecosystems.

Overall, climate change would very adversely affect crop production, marine and animal life. Along with the effects of climate change on agriculture, other evolving factors that affect agricultural production, such as, changes in farming systems and technology need to be addressed. Table 3.1 summaries the effects of climate change on several aspects of human life and the ecosystem.

What Are the Projected Impacts in Asia?

South Asia will become the worst affected by climate change inasmuch as its food security is concerned, as nearly 1.6 billion persons will become vulnerable. The crop model predictions indicate that yield level of rice and wheat will drop by 17 and 50% from 2000 to 2050. It is also projected that the availability of average calories per person in 2050 will be reduced by 15% from the current level. It will increase the number of malnourished children and stunting in children below age 5.

Direct Impacts of Climate Change on Agriculture

Change in weather patterns, as a result of climate change, leads to lowered crop production, lowered nutrition, excessive pest and disease attack on crops, and water scarcity.

Table 3.1 Effects of climate change on several aspects of human life and the ecosystem

Parameters	Climate change consequences	
Ocean	Sea level	Melting ice sheets and glaciers cause sea level rise consequently inundating low-lying areas. Islands, coastal cities and farmland Near sea very vulnerable
	Acidification	Dissolved CO_2 renders sea water more acidic harming sea life
	Marine oxygen concentration	Less dissolved O_2 in sea water adversely affects marine life leading to disequilibrium of species
	Temperature increase	Plant types, corals and animals that can live in each location will change and aquatic species migrate to warmer regions
Natural resources	Freshwater	Fresh water available for drinking and irrigation declines. Rivers, springs etc. dry up
	Vegetation	Shifting of agro-ecological zones and plants and trees move pole ward up the mountain slopes
	Wildlife	Areas in which animals live may change and some animals and plants may move closer to cooler places. Some species expand, others shrink
	Biodiversity	Both plant and animal biodiversity will decline
	Extinctions	A quarter of global species may become extinct and their habitats will decline
	Ice sheets	Ongoing melting of ice will accelerate in Antarctica and Greenland
	Glaciers	Will recede, disappear and disrupt quantity and timing of river flows
Society and economy	Migration	Changes will compel humanity to move from one place to another in search of newer livelihoods. Those who moved to coastal areas will further be compelled to move because of rise in sea level
	Conflict	Competition of livelihood resources, such as, food, water and fuel will escalate
	Poverty	Some people are less able to adapt to climate change than others, thus, will make them more vulnerable
	Economy	Both business and economy will be hugely impacted by climate change, which will cost the economy of nations from 5 to 20% in the gross domestic product (GDP)
Human health	Diseases	Diseases, such as, malaria, diarrhoea and dengue fever will escalate
	Heat waves	Periods of extreme heat will increase the frequency of putting more people to greater risks of all kinds, health, monetary and psychological

Effect of Rising Temperature

Rising temperature may alter the acreage for farming, reduce crop yields in tropical areas. It is predicted that moderate increase in temperature m about 1–2 °C, may affect crop yields negatively. However, higher winter temperatures and more number of frost-free days in some regions would result in the lengthening of the growing season (Kulshrestha 2011).

Effect of heat waves: Both growth and development of the plant will be adversely affected, ultimately reflecting in lower crop yields.

Heat stress on livestock: Heat stress adversely affects animal biology resulting in lowered milk yield, in cows, poor egg production in poultry, and lowered meat production in other animals like pigs. Different species of livestock differ in their capacity to withstand heat stress.

Changes in rainfall pattern: These cause both droughts and floods which adversely affect crop production. In drought, grazing lands ae adversely affected, affecting livestock productivity. Recurrent droughts compel small land holders to abandon agriculture and go in search of other livelihoods.

Carbon dioxide fertilization: Photosynthetic rate is enhanced through elevated CO_2 production.

Reduced nutritional value of crops: CO_2 levels in the atmosphere affect the level of protein and nutritional quality of crops. It is more pronounced in C_3 crops, such as, rice, wheat, oat, etc. Food crops can show reduced level of protein, iron and zinc, among important micronutrients. These will, inevitably, adversely affect the health of humans consuming these pant products from these crops.

What Are the Indirect Impacts?

Prevalence of pest and diseases: Pests, both diseases and insects, attack crop plants and the global warming caused by anthropogenic factors will enhance their prevalence and infestations. Warm and humid weather is always very conducive for the multiplication of these pests. Climate change will invariably increase the pest population and the activity of desert locusts, fall army worms, weeds, invasive species and plant pathogens. It has been observed that cold night temperatures, especially in winter months, would kill insects, bacteria and fungi.

Food security, under nutrition and food process: It is invariably observed that climate change enhances loss of food and its production. This will have a cascading affect on food prices. Food insecurity results due to food unavailability and loss of purchasing power, consequently under nutrition results. According to IPCC's (Intergovernmental Panel on Climate Control, established in 1988 as per the mandate of the United Nations) Special Report, on Climate Change and Land, food prices will rise by at least 80% by 2050.

Loss of agricultural land: Agricultural land adjacent to sea shores will be inundated and washed away by the impact of rising sea levels, consequent to enhanced sea temperature, which will lead to rising sea levels. Soil erosion, submergence of shore lines, salinity will result.

Reduced Irrigation

Water runoff from melting glaciers may be depleted from climate change. Such runoff will lead to water depletion, consequently adversely impacting water availability for both human consumption and agriculture.

Erosion and soil fertility: Warmer atmospheric temperature will impact hydrological cycles, causing excess rainfall. This will result in soil erosion. Warmer climates could lead to increase in the population of pathogenic bacteria in soil.

Change in flowering pattern: Warmer climate can induce earlier flowering and earlier blooms will threaten plant survival and reproduction.

Impacts on surface levels of ozone: Surface ozone layer caused through methane emission will act as an air pollutant, which will reduce physiological activity thereby adversely impacting plant production.

Potential effects of climate events on agriculture: Increased atmospheric temperature damages soil physical structure, leading to soil erosion and invariably adversely affect soil fertility. These effects will be more visible in lower than in higher altitudes. This situation is, in particular, harmful to developing countries, as compared to the developed ones. Depending on the prevalent scenario, researchers have estimated that crop yields could decrease by 50% in wheat, 45% in maize, 30% in rice and up to 60% in soybean. Table 3.2 depicts the projected impacts of climate change o crop yields.

Impacts on livestock: Both productivity and health of livestock adversely impacted. Reduced fodder availability will lead to lower milk yields. While pasture productivity is increased by elevated CO_2 concentration, there's a lowering of its quality. Drought and waterlogging create problems of drinking water availability for animals.

Fisheries and aquaculture: Reproduction timing and migration of fishes adversely affected by temperature changes. Capture fisheries and aquaculture will suffer from rising water temperatures and oxygen deficiencies. In oceans, sharp climate variations will change the habitats of aquatic animals, as well as composition of fish species.

Table 3.2 Projected climate impacts on crop yields

Climate change	Crop impact
Cold periods becoming warmer and shorter; over most land areas, days and nights becoming hotter	Increased crop yields in colder environments; increased outbreak newer disease pathogens and insects; potential adverse impacts on crop production
Heavy precipitation incidents increasing in frequency over most areas very likely	Damage to standing crops; soil erosion, waterlogging, rendering field cultivation very difficult
Land degradation and soil erosion; loss of arable land, loss of crop yield due to damage	Extensive crop damage
Extremely high sea levels increase tsunami occurrence	Salinization of irrigation water, estuaries, and fresh water systems; arable land loss
Drought affected areas increase	Land degradation and soil erosion; loss of arable land; lower yields due to crop damage

Forestry: Weather aberration limits the capacity of forests to produce goods and services as a result of which forest dwellers are most adversely affected. Extreme climate events may drastically alter the biodiversity of forest eco systems.

References

IPCC (Intergovernmental Panel on Climate Change). 2007. Climate Change: The Physical Science Basis. Extracts from the IV Assessment Report. Survey of the Environment 2007, The Hindu, pp. 147–155

Kulshrestha, S.N. 2011. Climate Change, Prairie Agriculture and Prairie Economy: The new normal. Canadian Journal of Agricultural Economics. 59(1), 19–44. doi: 10. 1111/j.I744-7976.2010.01211.x

Nair, K.P.P. 2019. Intelligent Soil Management for Sustainable Agriculture: The Nutrient Buffer Power Concept. Springer Nature Switzerland AG 2019

Nair. K.P.P. 2023. Biodiversity in Agriculture Sustainability of Soil, Soil Fauna and Soil Flora. Springer Nature Switzerland AG 2023

Nair, K.P.P. 2024. India's Tryst with Green Revolution - A Critical Aanysis. Agricutural Researfch Updates Vol 47, 117–145 Nova Science Publishers Inc. New York, USA

Paul Erhlich on climate change 1968. Backseat driving. Retrieved November 17, 2005, pages 51–52 of The Population Bomb, 1968, available from Schmidt Brian (July 1, 2005)

Chapter 4
Challenges of Climate Change for Sustainable Food Security

Agriculture and Food Security: Food security refers to a situation when people gain physical access to sufficient food to meet the dietary requirements on a daily basis. A household can have food security when all the members lead a hunger fee life. India has been unable to achieve this target, despite the much talked about "green revolution". It is estimated that nearly 800 million persons go to be bed each day with a hungry stomach in the world. Food security exists when everyone in the world, at all times, has physical and economic access to sufficient, safe and nutritious food which meets their dietary needs on a daily basis and food preferences for an active and healthy life (World Food Summit 1996). This definition gives rise to four dimensions of food security, as detailed below:

1. Food availability: Availability of sufficient quantity of food, of appropriate quality, for easy access is achieved through domestic production, distribution and also through exchanges. When domestic production falls short of the requirement of a country, it manages through imports. The available food should be appropriately distributed to all citizens of the country.
2. Accessibility of food: Physical, economic, and environmental accesses are very important to ensure food security of a country. The individuals should have enough resources to buy food, which is called purchasing power.
3. Utilization or food use: In order to meet the physiological needs, the individuals should have easy access to adequate diet, clean water, sanitation and healthcare. Nutrient content of food, food safety, and human health are the prerequisites for proper utilization of food.
4. Stability: This relates to all time food availability, access and utilization of food. To achieve this, a population or household or even an individual must have easy access to food at all times. At any time, they should not face the consequences of sudden shocks or cyclic events, such as seasonal food insecurity.

K. P. Nair, *Climate Resilient Sustainable Agriculture*,
https://doi.org/10.1007/978-3-031-84166-8_4

Table 4.1 Climate change impact on food security

Dimension	Impact of climate change
Availability	Climate change causes weather aberration non conducive for adequate food production. Declines in wild food availability and limits on small-scale horticultural food production, adversely nutritional status of a population. Climate change causes new patterns of pestilence, threatening food security
Access	Climate change adversely impacts income earning capacity severely impacting easy access to food, primarily due to economic factors. High food prices has a cascading effect on the nutritional status of a country's population
Utilization	Climate change adversely impacts the accessibility of certain foods of important nutritional value. This will lead to change in food habits, ultimately having a cascading effect on the health of nation
Stability	Droughts and floods are a particular threat to food security. Droughts caused by the El Niño and the tsunami are of particular importance. Price instability of food in the global market is a direct consequence of this

Agriculture has a major role to play in achieving food security of a nation. Insufficient food production and is easy access, climate change, natural disasters, conflicts, displacement, and population explosion, are among the many challenges faced by the community to achieve food security (Table 4.1).

Generally tsunamis arrive, not as giant breaking waves, but as a forceful rapid increase in water levels that results in violent flooding. However, when tsunami waves become extremely large in height, they savagely attack coastlines, causing devastating property damage and loss of life. El Niño is a climate pattern that describes the unusual warming of surface waters in the eastern equatorial Pacific Ocean. Trade winds and atmosphere are also impacted by El Niño.

Against the current population explosion, globally, it is estimated that the world will have close to 10 billion people by 2050. The current challenge facing agricultural scientists would be how best one can close the yield gap. As of today, there are many, both biotic and abiotic stresses that limit crop yields. Among the latter category, anthropogenic genic warming is an important one. Many of the crop varieties/cultivars, introduced since the green revolution, have hit their yield limits, and, the crop yields are either plateauing or declining, despite the recommended inputs, most importantly, fertilizers. These issues have been elaborately discussed by this author, in his blockbuster book, titled, "Intelligent Soil Management for Sustainable Agriculture: The Nutrient Buffer Power Concept" (Nair 2018). It is in this context that the utilization of "Crop Wild Relatives" assume critical importance, especially in combating anthropogenic warming. These issues have been discussed in great detail by this author in his another blockbuster book titled, "Combating Global Warming: The Role of Crop Wild Relatives for Food Security" (Nair 2019).

Climate change and livelihoods: The livelihood of a person refers to "means of securing the basic necessities of life". Livelihood comprises the capabilities, assets (including both material and social resources), and, activities required for a means of living. The five assets essential for a person's livelihood are, physical, financial, natural, social and human. A livelihood is sustainable when it can cope with and recover from the stresses and shocks, and maintain or enhance its capabilities and assets both at present and in the future, while not undermining the natural resource base (Carney 1998).

Livelihoods have relationship with climate change on all assets and adaptations, which included activities, such as, poverty reduction, improving access to resources, lowering resource and wealth inequities, improving education, infrastructure, institutional capacity and efficiency, and, finally, promoting knowledge acquisition. These five assets change over time and are different for different households and communities. Physical assets can be damaged or destroyed, financial losses can be incurred, natural assets can be degraded and social assets can be undermined. Climate change may cause several variations in food supply, along with vulnerabilities to flooding, fire, which can make livelihoods more vulnerable at certain times of the year. The following are the ways in which assets for a livelihood are also affected by climate change in different ways.

1. Natural capital: Natural resources like land, water, forest and biodiversity available to an individual influence the livelihoods. Climate change leads to increase in arid lands, causes land degradation and erosion, and water stress, while prolonged drought or waterlogging alter the physical, chemical and biological properties of the soil resources. Growth in forest ecosystems is reduced due to climate change. All of these changes threaten the livelihood security of mankind.
2. Physical capital: Basic infrastructure for transport, buildings, water management, energy, and communications and productive capitals (tools, machines etc.), which enable people to pursue their livelihoods. Both private and governmental agencies provide physical assets. Loss of physical assets have multiplier adverse effects. Floods, landslide and other physical events affect the physical capital.
3. Social capital: Social asses like group and networks are affected by climate change through excessive rains, cold and heat waves.
4. Human capital: Human skills, knowledge, good health etc. are influenced by climate change. Climate change is effective in different ways in men and women.
5. Financial capital: Monetary savings, pensions, social security payments etc., are adversely affected by climate change, because they have a cascading effect of human endeavor.

Exploitation of common property measures (CPRM) for livestock, fishing, fuel gathering and such other things, which have a direct impact on livelihoods are affected by climate change. Climate induced changes to resource flows (be they temporary, which reflect variability or structural, reflecting change) are affected by climate change.

Achieving Food Security Through Adaptation and Mitigation of Climate Change

Both adaptation and mitigation strategies are required, urgently, to adjust and contain the severity of climate change on all aspects. Adaptation refers to some measures of adjustments to decrease the vulnerability of production systems to climate change. Mitigation is the effort to reduce the adverse impacts of climate change. Many technologies have been developed to enhance crop production systems to enhance their abilities to adapt and mitigate the adverse effects of climate change. Adaptation strategies focus on disaster risk management and risk transfer mechanisms like livelihood systems, integrated crop-livestock and aquaculture-agriculture systems and crop insurances. Rao et al. (1989) in India found that technological innovations in conjunction with input management and adoption management practices result in stability of sorghum and millets cultivated during rainy season. Nair (2019) has proposed the very important idea of utilizing Crop Wild Relatives in containing anthropogenic warming.

Adapting to climate change: Adaptation to climate change includes good practices to strengthen their resilience to climate change and management of risks by required information and cop insurance coverage. It also seeks a new institutional and technological solution against adverse impacts of climate change.

Components of climate change adaptation: it is essential to adopt into an integrated approach to manage climate risks at various levels. The activities to promote adaptation and reduce vulnerability require following strategy element (given below) in bottom-up manner, as suggested by FAO.

1. Assessing current vulnerability: Assessment of natural and socio-economic conditions, current climate risks, local perception about climate risks and impacts, and documentation of livelihood profiles along with institutional framework would be required for the pilot sites. These information will be useful to understand the vulnerability to climate risk.
2. Assessing future climate risks: Assessment of the impact of climate change, agriculture scenario and local agro-meteorological data, which would form the basis for future climate risks in the concerned region.
3. Testing adaptation options: Identification and selection of viable options for adoption of farmer-friendly practices along with extension strategy. Community mobilization will have to be made to raise people's awareness to climate change.
4. Design adaptation strategy: Adaptation strategy for dissemination and replication of essential options will be prepared by raising awareness, economic feasibility studies and field-based demonstrations and applications.

Mitigation of climate change: Food and livelihood security can be protected by mitigating climate change. Conversion of forest land to pasture or cropland, and intensive crop and livestock production practices have contributed much to containment of green house gas emissions (GHGs). Therefore, appropriate practices to reduce CO_2 emissions and also emissions of other GHGs have to be adopted. Some mitigative aspects are discussed below:

1. Reducing CO_2 emissions from agriculture and forestry: Forest fires and stubble burning practices have to be contained. UNFCCC and Kyoto Protocol recognize the potential role of forests to provide a variety of adaptive ecosystem services in addition to mitigating climate change through carbon sequestration. Afforestation, agro-forestry and integrated farming systems with tree component are sound practices to contain CO_2 emissions.
2. Reduction of methane (CH_4) and nitrous oxide (N_2O) from agricultural practices. The enteric fermentation of digested fibers by large ruminants (such as cows) is the important source of methane emission. Rice cultivation under flooded condition is another source. Unbridled and excessive nitrogen fertilizers, like urea, is the principal source of nitrous oxide through microbial decomposition. Controlled and sustainable fertilizer management (Nair 2018) can reduce nitrous oxide emission. Grazing and feed management in livestock farming is another option in containing methane emission.
3. Conservation agriculture: This involves several types of cultivation, such as, zero tillage, minimum tillage, surface mulching, cover cropping etc. In conservation agriculture, mechanical intervention, such as tillage operation is reduced to the minimum required. Further, fertilizer management is practiced to the optimum. In this connection, "The Nutrient Buffer Power Concept", a revolutionary soil/ fertilizer management technique developed by this author (Nair 2018), will be of immense utility. Crop yield fluctuations when conservation agriculture is practiced is minimal (Tebrugge and Bohmsen 1998; Derpsch 2005).
4. Biodiversity conservation: Generally diverse populations and species-rich ecosystems can adapt well to climate change. FAO promotes the use of indigenous and locally adapted plant and animal diversity, and, the selection and multiplication crop varieties and autochthonous races which are adapted or resistant to adverse conditions. Breeding plants and animals for tolerance to abiotic stresses, such as, drought, heat, salinity and flooding will also become increasingly important. All these aspects are elaborately discussed by this author in his recent book (Nair 2023) titled "Biodiversity in Agriculture—Sustainability of soil, soil fauna and soil flora".
5. Carbon sequestration: The method of enhancing carbon storage in terrestrial ecosystems is termed carbon sequestration. Examples, such as, planting trees, which act as "carbon sinks", is a very important practice to contain adverse climate related impacts. Management practices for carbon sequestration have additional benefits, such as, increased plant root mass, through soil organic matter accretion, beneficial nutrient and soil water retentive capacity.

Climate Solution Farming

The objective of climate solution farming is to find out ways and means which can significantly lower green house gas (GHGs) emission, store additional carbon from the atmosphere in the soil, reduce vulnerability to the adverse impacts of climate change. Instead of excessively using synthetic fertilizers, such as urea, which can have a very deleterious impact on anthropogenic warming (Nair 2019), use of organic manures, such as, farm yard manure (FYM), vermicompost, use of green manure plants, such as Sunhemp (*Crotalaria juncea*) when, *Gliricidia maculata* etc., which build up soil organic matter, with appropriate crop rotation practices will enhance build up of carbon sequestration.

Although the above-mentioned practices may not fully replace the use of synthetic fertilizers, especially with nitrogen-dependent crop plants, they can significantly reduce the quantum of synthetic fertilizers used in crop production. Another possible solution is through perennial farming systems, such as, agro-forestry, tree planting and fruit tree cultivation. These crops have multiple benefits, such as, significantly reducing the need for tillage; deeper root systems that both protect and build soil, better drought tolerance, and lower fertilizer (synthetic) and chemical pesticides, insecticides and herbicide requirements. Adding more carbon to the soil can help reduce the damaging effects of these GHGs in the atmosphere and mitigate some of the deleterious impacts on crop production of our fast changing climate pattern. No-till farming, crop rotation and cover cropping are very welcome options.

References

Carney, D. (Ed.) Sustainable rural livelihoods. London, DFID, 1998

Derpsch, R. 2005. The extent of CA adoption worldwide: Implications and impacts. Paper presented at III World Congress on Conservation Agriculture, Nairobi

Nair, K.P.P. 2018. Intelligent Soil Management for Sustainable Agriculture: The Nutrient Buffer Power Concept. Springer Nature AG Switzerland 2018

Nair, K.P.P. 2019. Combating Global Warming: The Role of Crop Wild Relatives for Food Security. Springer Nature Switzerland AG 2019

Nair, K.P.P. 2023. Biodiversity in Agriculture Sustainability of Soil, Soil Fauna and Soil Flora. Springer Nature Switzerland AG 2023

Rao, N.G.P., Rao, G.R.K., and Acharya, H.S. 1989. Yield Stability of Sorghum and Millet Across Climates: In: Climate and Food Security International Symposium on Climate Variability and Food Security in Developing Countries, 5–9 February, 1987, Manila, Philippines: International Rice Research Institute and Washington, DC, American Association for the Advancement of Science

Tebrugge, F., and Bohmsen, A. 1998. Farmers' and Experts' opinion on no tillage. Results of a Europe wide Survey. In the Proceedings of the Conference of Agricultural Engineering 1998. Dusseldorf, Germany. Association of German Engineers, Society of Agricultural Engineering (VDI-MEG)

World Food Summit, 1996. Rome Declaration on World Food Security, Rome, Italy, FAO www.europaworld.org/issue/conservation51001.htm

Chapter 5
Climate Resilient Agriculture

Climatic factors, such as, increase in atmospheric temperature, atmospheric concentration of CO_2, moisture and many other factors directly affect crop production. These aspects have been elaborately discussed in the previous chapters. Climate change causes weather variability which adversely affect food production which will have a cascading effect on food prices. This makes agriculture a vulnerable avocation. It is now realized that climate resilient agriculture (CRA) will be imperative to reduce the vulnerability of agriculture to climate change.

Climate Resilient Agriculture (CRA)

CRA implies the incorporation of adaptation, mitigation and other practices in agriculture to address the adverse impacts of climate change. It includes judicious and better management of natural resources through adoption of new technologies and better and more efficient management of natural resources through adoption of new technologies and practices. CRA can address the current adverse impacts of climate change and help sustain agricultural production from the local to the global level, in a sustainable manner.

CRA is an integrated approach which aims at the following three objectives:

1. Sustainably increase agricultural productivity, to support equitable increases in farm incomes.
2. Adapting, and building resilience of agricultural and food security systems.
3. Reducing emissions of green house gases (GHGs) from agriculture, including crops, livestock and fisheries to the maximum extent possible.

CRA practices: Various practices to be adopted for climate resilient agriculture can be grouped under agriculture sub-sectors following the guidelines of FAO (2021).

K. P. Nair, *Climate Resilient Sustainable Agriculture*,
https://doi.org/10.1007/978-3-031-84166-8_5

Crops and cropping systems: The crops grown under different situations are usually affected by extensive heat, strong wind, cold waves, hailstorms, drought, flood and waterlogging, land degradation, GHGs emission, salinization and pest attacks caused by diseases and insects. Various methods/practices against each of such disruptions are outlined below with possible benefits.

Extreme heat: Use of heat tolerant crops, such as millets, for example, sorghum (*Sorghum vulgare*), Pearl millet or bajra (*Pennisetum typhoideum* or *Pennisetum glaucum*) etc., and other short duration cops and adjustment of time of sowing to avoid heat stress, can reduce the adverse impact of heat and result in increased crop yield.

Drought: The resilient practices include adoption of agro-forestry and alternate cropping systems, selection of various drought tolerant crops and their varieties, adjusting planting seasons to minimize exposure to heat stress, construction of terrace, mini soil and water conservation measures, such as, half-moon terrace, staggered trenches, farm ponds, broad bed and furrow system (BF), provision of life saving system from water harvesting structures, rain water harvesting, timely weeding, cover cropping, mixed and intercropping etc., which reduce the adverse impact of drought.

Cold, frost and hail: Protection from frost can be made by planting tree rows and frost protection irrigation. Soil mulching increases surface temperature and canopy cover can provide protection against frost damages by reducing net radiation losses. Early warning and adjusting time of planting will reduce the adverse impact of cold and frost.

Flood and water logging: Provision of drainage facilities, raised bed system of cultivation, field dredging, Agronomic measures, such as, sub-soiling, application of organic manures, and provision of bio-drainage can reduce the adverse impact of flooding and waterlogging. Cultivating flood-resistant/tolerant varieties of crops can minimize the adverse impacts of waterlogging and flooding.

Land degradation: Crop rotation, cover cropping, cultivation of legume crops, contour farming, strip copping, mulching, minimum tillage, application of organic manures, cultivation of green manure crops, such as, Sunhemp (*Crotalaria juncea*), Gliricidia (*Gliricidia maculata*), early sowing of crops to encourage a vigorous root system and adoption of soil conservation measures, can protect land degradation and make the crops to resist/tolerate climate change.

Reducing emissions of GHGs (green house gases): Adoption of wetting, drying technology in rice cultivation, direct seeding, system of rice intensification (SRI), and, atmospheric carbon fixation in C_3, C_4, precision agriculture, specific nutrient management of plant nutrients by using green seeker and leaf color chart (LCC), integrated nutrient management, and increasing use efficiency etc., are proven technologies to reduce emission of GHGs. A separate, following chapter, will discuss the merits of a revolutionary soil management developed technique by this author, currently globally known as "The Nutrient Buffer Power Concept" (Nair 2019a), in reducing adverse impacts of climate change.

Salinization: Promotion halophyte crops (Halophytes are defined as plants that naturally inhabit saline environments and benefit from having substantial amounts of salt in the growth media. Halophytes grow in a wide variety of saline habitats, from coastal regions, salt marshes and mudflats, to inland deserts, salt flats and steppes), use of salinity tolerant crops, sub soil drainage soil amendments, such as gypsum to correct soil pH, measures against vertical movement of saline water, construction of barriers to restrict inundation of sea water, seed bed management and field grading and irrigation system management, are resilient practices to contain climate change.

Management of crop pests and diseases: Integrated pest management involving host plant resistance/tolerance, cultural control, physical, mechanical and biological pest control measures are required to limit the attack of insect pests and diseases, within the economic injury level. The crop management practices should be so designed that it would encourage the activities of natural enemies of the pests, both insects and diseases, in the field. Biological control employing predators, parasites, nuclear polyhedrosis virus (NPV), botanical pesticides, such as neem cake, also *karanj*, various traps, such as, pheromone traps, yellow traps, sticky traps, and regular pet surveillance in crop fields are very effective against pest and disease attack.

Livestock system: The production of livestock is affected by climate change as a result of which, production of milk, egg and meat is adversely affected and the prices of these commodities are on the increase. The climate resilient practices to reduce the effect of adverse climate on livestock are detailed below.

Drought and heat: In order to save livestock from drought and heat, measures, such provision of drinking water, reduction of overgrazing, introduction of livestock breed which are heat stress tolerant, provision of shelter, integrating agriculture and forestry with livestock production, provision of shelter belts, biodiversity conservation and livestock mortality reduction, by reducing exposure to these abiotic stresses must be adopted.

Land degradation: Rotational grazing, fodder management, soil management, herd size reduction, introduction of silvi-pastoral farming, livestock based integrated farming etc., can be adopted to reduce land degradation.

Emission of GHGs: Nutrition improvement, manure management, diet supplementation, biogas production etc. can ensure this.

Animal diseases: Control of flies, mosquitoes etc., in animal sheds can control animal diseases.

Forest System

Degradation of forest land: A stock-taking exercise has to be made before forest planning. Forest inventory is required to assess the forest growing conditions, forest sock resources, forest hazards etc., before forest planning. Sustainable management of forest ecosystem, biodiversity conservation, control of grazing, controlled timer harvesting, control of forest fire, water conservation, conservation of top soil, afforestation, etc., are all viable options.

Emission of GHGs (green house gases): Reforestation and afforestation for carbon sequestration control of forest fire by stone fencing, restricting shifting cultivation and trash burning, nutrient management, increasing soil moisture etc., are required to limit GHGs.

Drought: Sustainable forest management, enhancing biodiversity to enhance resilience to dry conditions, mixed plantation, avoiding clear-cut logging and preserving forest biome, soil and water management interventions may reduce impact of drought.

Pests and diseases: Pest monitoring, cultural management, such as, pruning, planting resilient crop varieties, removal of pests from their habitats by introducing mechanical traps etc. are required to control pests and diseases.

Fisheries and aquaculture: Limited fish catch from seas, restoring mangrove plantations, reducing coastal soil erosion, reducing ocean acidification disaster management with early warning systems, construction of seawalls and flood reservoirs or buffer zones via afforestation or reforestation etc., are to be practiced.

Climate Resilient Agriculture Through Adaptation and Mitigation Strategies

National Innovations in Climate Resilient Agriculture (NICRA) was launched by the Indian Council of Agricultural Research, New Delhi, India, in 2011. The basic objectives of the flagship program were:

1. To undertake strategic and applied research on climate adaptation and mitigation.
2. To validate, demonstrate and assess, the impact of climate resilient technologies on farmers' fields.
3. To strengthen the capacity of scientists, farmers, and other stake holders on climate resilient agriculture.

The following village level interventions were made in the project area:

A. Building resilience in soil: Building soil carbon, controlling soil erosion, enhancing water holding capacity, balanced soil nutrition etc.

Table 5.1 The different crop varieties evolved by the Indian Council of Agricultural Research, New Delhi, India

Location	Crop	Variety	(Duration in days)
Varanasi	Rice	Vandana	93
	Pigeon pea	T.21	176
Phulbani	Pigeon pea	T.21	160
Arjia	Maize	Surya	70–75
Anantapur	Groundnut	Vemana	105–110
Indore	Soybean	JS-90-41	87–98
Rewa	Rice	Kalinga-3	90
Akola	Cotton	AKH-081	150–160
Bjapur	Sunflower	KBSH-1	90–95
Solapur	Sorghum	Mauli	105–110
Hisar	Pearl millet	HHB-67	60–62
Bengaluru	Finger millet	GPU-26	90–105

Source CRIDA, Hyderabad

B. Adapted cultivars and cropping systems: Using improved early duration, heat, drought and flood tolerant crop varieties. In this connection the idea of utilizing "Crop Wild Relatives" (Nair 2019b) will also be very beneficial.

   Table 5.1 details the different crop varieties, with their duration, developed in different Indian regions.

C. Water saving technologies: Technologies, such as, zero tillage, direct seeded rice help save water and control adverse effects of green house gases.

D. Farm machinery (custom hiring) centers: Improving access to farm machinery which will help farm mechanization, through community—managed hiring systems of farm machinery.

E. Crop contingency plan: Indian Council of Agricultural Research (ICAR), New Delhi, India/Centra Research Institute for Dryland Agriculture (CRIDA), Hyderabad, India, have prepared contingency plans for 400 districts in India to cope with climate change.

F. Livestock and fisheries intervention: Use of community land for the production of fodder, storage etc. at the level of villages, feed supplements to reduce nutrient deficiency in livestock, disease preventive vaccination, livestock shelters to reduce heat/drought shock to animals and fishery management to enhance fish production.

G. Weather-based agro-advisories: Installation of automated weather stations at KVKs (***Krishi Vigyan Kendras***—Centers for Agricultural Research at the village level), mini weather recording equipment at project sites, to help offer proper advisories to farmers.

H. Institutional interventions: Initiation of establishing seed banks, fodder banks, custom hiring centers, collective marketing faculties for farm produce, and weather index based crop insurance facilities to improve farming effectiveness on the whole.

I. Village Climate Risk Management Committees (VCRMC): Promotion of farmers' participation through VCRMC constituted taking all types of farmers, Including women and the landless, with the approval of ***Gram Sabha*** (Village Group) for taking all decisions on climate resilient agriculture.

Key Technologies/Strategies for Climate Resilient Agriculture

The climate resilient agricultural technologies can be grouped as follows for convenience of planning and implementation:

1. Building soil resilience: Soil health and soil fertility management are two very important factors in building soil resilience. Building up soil carbon, control of soil erosion and soil degradation, increasing soil water holding capacity, balanced soil fertilization as per crop requirement are some of the factors. More than 50% of a crop's/plant's performance is dependent on the "nutrient factor". This can be managed only through precise soil testing and calibrating for the right amount of fertilizer to be applied to the soil. These aspects are elaborately discussed in the book authored by this writer, titled "Intelligent Soil Management for Sustainable Agriculture: The Nutrient Buffer Power Concept" (Nair 2019a), where this writer elaborately discusses a revolutionary soil management technique, now globally known as, "**The Nutrient Buffer Power Concept**".

   Site-specific nutrient management (SSNM) is a set of location-specific scientific principles for optimally supplying rice with essential plant nutrients based on specific conditions. The principles of "**The Nutrient Buffer Power Concept**" of this author (Nair 2019a), can apply equally well for rice, though he had not considered rice as a test crop, but, tested the validity of the concept in wheat, maize, white clover, black pepper and cardamom, the latter two among perennial cops. Specific nutrient management using leaf color chart (LCC) and GreenSeeker, also will be useful.

   Integrated nutrient management deals with the application of organic and inorganic fertilizers, in addition to farm yard manure (FYM), vermicompost, growing legumes in crop rotation, and using crop residues for sustaining soil health for the long term.

   Conservation agriculture is useful to enhance resource use efficiency, provide economic benefits to the farmer, in combating and containing climate stress. It improves soil structure and protects the soil against soil erosion and degradation.

   Zero tillage or no tillage is designed to generate higher crop yields, while minimizing soil erosion and reduces input costs. The zero tillage practice reduces CO_2 emissions and saves water usage up to 25% compared to conventional methods. The largest CO_2 emission reduction is achieved through reduced diesel fuel consumption during ploughing operations. The zero tillage method can allow farmers to sow/grow wheat immediately after the harvest of the rice crop so that the crop escapes the terminal heat stress. Bed planting with wheat residues saves on water on irrigation usage.

Laser Land Levelling (LLL) is a laser-guided technology used to level crop fields by removing soil from the high points of the field and depositing it in low points of the field. LLL technology improves crop establishment and enables the crop to mature uniformly. In the Indo-Gangetic Plains of India, crop yields were found to have increased by 8% in both rice and wheat crops. In Cambodia rice yield improved by 24%. LLL also reduces GHGs emission. In a level field, water is distributed evenly, thus reducing the amount of time and volume of water used for irrigation.

2. Adapted cultivars and cropping systems: Growing adapted cultivars of crops or breeds tolerant to abiotic stress and adoption of diversified cropping system can reduce the risks of crop failure due to climate change. Crop management practices, such as, altering planting dates, choice of short duration varieties with tolerance to stressful conditions is helpful in combating climate change. Muller and Elliott (2015) have found that adaptive changes in crop management practices have the potential to enhance yield by 7–15%. Improved and early duration crop varieties adapted to climate stress must be promoted and their seed production encouraged. Patterns of drought may need various forms of adoptive measures.

   Local varieties resistant to abiotic stresses can also be identified and used as climate resistant cultivars, along with the promotion of seed banks, seed villages and participatory variety selection. A sequential cropping system involving both shallow and deep rooted crops with soil enriching capacity is preferred. Crop diversification instead of monoculture, accommodating a variety of crops, can lessen the risk abiotic risks from climate change.
3. Rain water harvesting and recycling: Many rainwater harvesting and conservation technologies, such as, construction of farm ponds, storage structures, restoration of old water harvesting structures, ponds for recharging ground water etc. are encouraged to be installed.
4. Water management: Adaptation measures, such as, water harvesting and storage, access to irrigation, improved irrigation technologies, as well as agronomic practices can enhance water retention capacity of the soil. Water-smart technologies like furrow-irrigated raised seed bed, micro irrigation, cover crop method etc. are encouraged for use. Alternate wetting and drying (AWD) is a simple and inexpensive method of reducing water usage in rice production by as much as 30%. This enables farmers to cut down on cost of production. AWD involves periodic draining of the field to a certain threshold, usually 15 cm below the soil surface and reflooding. A perforated tube placed below the soil surface which helps the farmer to schedule his irrigation properly.
5. Farm mechanization: Farm mechanization is very critical for timely field operations and crop planting, besides, minimizing drudgery. Community managed custom hiring centers having sowing, planting machines, chisel plow etc. come in handy in this endeavor. Dry seeded rice (DSR) has become an attractive option for the rice farmer in India and for this 2–4 wheel tractor/power tiller

drawn seed drills come in handy. DSR technique can be successfully used both by small-scale farmers as well as large-scale ones.

6. Landscape management: Many ecosystem services, relevant to agricultural production, such as, pollination, pest and disease resistance, watershed protection and soil erosion control, occur over landscape scales. These services are directly related to resilience of agricultural livelihoods through their impacts on reducing environmental risks and improving coping up capacities (FAO 2007; McCarthy et al. 2010). Watershed protection and management, fire management, soil erosion control, coastal zone management, pest and disease control required for landscape management. Landscape-level adaptation will require appropriate institutions and policies to improve coping up capacities of human communities.
7. Livestock and fisheries intervention: A range of adaptation options, such as, feeding/housing systems, production system and breeding can lead to climate resilience. Livestock's adaptive capacity depends on the production system, including choice of species and breeds, availability/adaptability of alternative feed resources, the accessibility of animals (health/extension services), the type/efficiency of response to outbreaks of diseases (surveillance, compensation in case of casualties etc.) and the household wealth status (ICEM 2013).
8. Tolerant breeds in livestock and poultry. Indigenous breeds have unique characteristics which are adapted to very specific ecosystems across the world. The very small-statured "***Vechur***" cow of Kerala State, India, is a classical example. These unique characteristics are, resistance to drought, thermoregulation ability, ability to walk long distances etc. The livestock breeds may not be high milk and meat producers, but, highly adaptive to unpredictable environmental conditions and have low resource footprints.
9. Feed management: Improvement in feeding systems as an adaptation measure can indirectly improve the efficiency of livestock production. Some feeding methods include, altering feeding time or agroforestry species in the animal diet and training producers in production and preservation of feed for different agro-ecological zones, frequency of feeding and modification of diet composition, including agroforestry species in the animal diet and training producers in production and conservation of feed for different agro-ecological zones. Adaptation options to declining or variable yields in terms of fisheries technologies and management should be assessed to avoid overexploitation of fisheries or adversely impacting habitats.
10. Diversified forest ecosystems: These are more resilient and they are able to cope up with climate stress and enable recovery. Practices such as integrated pest management, disease control, forest fire management, employment of reduced impact of logging in productive forests, limitation of gathering on non-wood forest products at sustainable levels, and forest law enforcement are found effective in developing a healthy forest ecosystem.

11. Integrated farming: Integrating different types of production systems (crop, forest, fish and livestock) can reduce the risk of production loss under climate stress. In some regions, switching from crop to mixed crop-livestock combination will be a key adaptation strategy (Jones and Thornton 2009).
12. Managing genetic resources: Crops, livestock, forest trees and aquatic organisms which can survive and produce in future climate patterns will be essential in future production systems.
13. Agro-advisory: The development of early warning systems is essential to strengthen proactive decision making at all levels. Analysis of meteorological data, together with information on phenology, soil and agricultural statistics, also allows the provision of near real-time dependable scientific information about crop forecast so that timely interventions can be planned and put in place. Response farming is an integrative approach which could called "farming with advisories taken from technocrats depending on local weather conditions".
14. Diversification of cropping on "on-farm" and "off-farm": Non-agriculture based livelihoods can build resilience among agricultural populations due to projected population growth patterns, as well as, potential climate impacts. Increasing number of crop varieties grown, species and animal breed, mixed-farming systems, such as crops/livestock, crop/fish or processing products on on-farm and getting a non-agricultural job, as off-farm jo, are some of the possibilities (Thornton and Herrero 2014).
15. National program for climate change adaptation in India: The convergence of various policy programs and sectoral plans has been undertaken, by the Government of India to ensure effective synergy and effective utilization of available resources. "The ***Pradhan Mantri Krishi Sinchayee Yojana***" is one such. Green India Mission is another. These models are taken forwarded to SAARC (South Asian Association for Regional Cooperation) countries for adaptation.

References

FAO 2007. The State of Food and Agriculture 2007. Paying farmers for agricultural services. FAO, Rome

FAO 2021. Climate Resilient Practices: Typology and Guiding Material for Risk Screening. FAO, Rome

ICEM (International Centre for Environmental Management) 2013

Jones, P.G., and Thornton, P.K. 2009. Croppers to livestock keepers: livelihood transitions to 2050 in Africa due to climate change. Environmental Science & Policy 12(4), 427–437

McCarthy, M., Best, M., Betts. R. 2010. Climate change in cities due to global warming and urban effects. Geophysical Research Letters 37(9)

Muller, C., and Elliott, J. 2015. The Global Gridded Crop Model intercomparison: approaches, insights an caveats for modelling climate change impacts on agriculture at the global scale. In: Climate Change and Food Systems. Global assessments and implications for food security and trade (A. Elbehri, Ed.). FAO, Rome

Nair, K.P.P. 2019a. Intelligent Soil Management for Sustainable Agriculture: The Nutrient Buffer Power Concept. Springer Nature Switzerland AG 2019

Nair, K.P.P. 2019b. Utilizing Crop Wild Relatives to Combat Global Warming. Advances in Agronomy, volume 153, 175–258

Thornton, P.K., and Herrero, M. 2014. Climate change adaptation in mixed crop-livestock systems in developing countries. Global Food Security 3(2), 99–107

USAID Mekong ARCC climate change impact and adaptation on livestock. Prepared for the United States Agency for International Development (USAID) by ICEM

Chapter 6
Fundamentals of Climate Resilient Agriculture

Introduction on Climate Resilient Agriculture (CRA)

Extreme weather conditions, such as, very high ambient temperature (El Nino), torrential rainfall bordering on flooding (La Nina), excessive drought, tsunami, hurricane, all of which, will have a very profound effect on global farming, which will be, invariably, very deleterious to production. This, in turn, will adversely affect global food security. Some of these are discussed in the book of this author "Combating Global Warming: The Role of Crop Wild Relatives for Food Security" (Nair 2019b). Emission of greenhouse gases (GHGs) is an additional and very important source of climate change. Climate change will, additionally, adversely affect food prices, especially for the principal staples like, wheat, rice, corn and soybean (IFRI 2009).

Climate-resilient agriculture is an innovative approach which can make the agriculture sectors more productive and sustainable with better ability to contributing towards climate change adaptation and mitigation. FAO defined climate-resilient agriculture (CRA) as the three dimensions of sustainable development (economic, social and environmental) jointly addressing food security and climate induced challenges. Agriculture that sustainably increases productivity and resilience to climate change reduces/removes greenhouse gases (GHGs), a mitigatory effect, and enhances achievement of national food security and developmental goals (FAO 2011).

The World Bank Group defines climate-smart agriculture as an integrated approach to managing cropland, livestock, forests and fisheries that address the interlinked challenges of food security and climate change. Thus, climate-smart agriculture is regarded as a new and original approach to support global farmers to achieve sustainable agricultural production and land management with the ability of the sector to adapt to and mitigate the adverse effects of climate change. CSA is a technique to improve the scientific rule and investment setting to attain sustainable agricultural progress to ensure food availability under climate change (FAO 2012). It can ensure food availability and conservation of natural resources. It relies on

K. P. Nair, *Climate Resilient Sustainable Agriculture*,
https://doi.org/10.1007/978-3-031-84166-8_6

Table 6.1 Indicators of CRA as notified by World Bank in 2016

Theme pillar	Sub theme	Indicators
Productivity	Crop system	Increases crop yield, reduces wind and water erosion risks, soil fertility enhancement
	Water use	Increase in irrigated land area
	Energy	Reduction in fossil fuel-based energy, enhancement of renewables like ethanol for locomotion
Resilience	Robustness	Enhancement in production stability, income diversification promotion, incorporation of site-specific knowledge
	Cropping system	Drought resistance enhancement
	Livestock system	Drought resistance enhancement
Mitigation	Emission intensity	Reduction of GHGs emission
	Sequesters carbon	Enhancement of carbon sequestration

adaptative processes which involve management of climate-generated risks. CSA is based on the following five pillars:

1. Increasing farm income for the farmer
2. Increasing food productivity
3. Increasing food security
4. Adaptation and resilience to climate change
5. Reducing GHGs.

Table 6.1 details the indicators of Climate Resilient Agriculture, as per the World Bank, notified in 2016.

Crop Production and the Five Pillars of CRA

Conventional plant breeding and the application of DNA molecular tools have been evolved to improve crop yields under abiotic stresses like drought, soil salinity etc. The above described five pillars come upon this connection. Adaptation measure like changing planting times and planting depth to suit seasonal weather aberrations can reduce/mitigate the risk of crop failure. Mitigation strategy largely seems from soil and water use management or agroforestry system. Use of alternative sources of energy, such as biogas, and solar pumps, instead of conventional fossil fuel-based energy will go a long way in reducing GHGs emission.

Rationale for Climate Resilient Agriculture

By 2050 we have to feed 8 billion human beings. This calls for 6% increase in current agricultural production. There are ever so many, both biotic and abiotic stresses, that crop production faces, as detailed above. CSA meets the challenge.

Objectives of Climate Resilient Agriculture

The following are the three principal objective of climate smart agriculture:

1. Sustainably maintain and increase crop productivity: In order to ensure food and nutrition security, our diets have to be balanced with carbohydrates, protein, fart, minerals, vitamins and micronutrients. Our future demand for food will include animal product, such as, milk, eggs, meat and fish. Hence, there is an urgent need to provide sustainable production of diversified food.
2. Adaptation and resilience: Anthropogenic warming, rising sea levels, extreme weather conditions due to climate change have affected agricultural production in many countries. In order to ensure resilience of agricultural systems, it is inevitable that the agricultural practices are adapted to climate change, and, resilient.
3. Mitigation: Rising level of GHGs causing climate change have to be reduced to acceptable levels. Currently agriculture and related sectors contribute a quarter of human-induced (anthropogenic) emissions. Hence, there is lot of potential to reduce emissions. And, to mitigate other detrimental effects of agriculture on the environment. Reducing GHGs and mobilizing soil carbon through sequestration are some of the important mitigative efforts/methods.

Approaches to Climate Resilient Agriculture

Climate Resilient Agriculture (CSA) is an approach which helps guide actions needed to transform and re-orient agricultural systems to effectively support the development and ensure food security in a challenging climate situation. CRA combines various sustainable methods to tackle the specific climate risks of a farming community. It is an integrated approach to managing different agricultural sectors to address interlinked challenges of food security and climate change.

The preliminary step is to assess the specific climate risks by using a variety of tools to assess the nature of the climate risk involved, and the vulnerability of the landscape, taking into consideration the local ecosystems and the specific crop in question. Finding the right combination to manage a specific climate challenge and to build resilience to future adverse climate impacts, makes the effort of climate-smart agriculture really climate "smart". New climate risks require changes in agricultural technologies and approaches to improve the food insecurity and poverty. Addressing the global challenges of climate changes requires generation, adaptation and use of new knowledge. This encompasses the need to link, more effectively and responsibly, to domestic and international markets (food, fiber, feed etc. and/or carbon pool in soil); reduce the vulnerability and enhance the voice of the rural poor; promote environmental conservation; build linkages between farmers and other agencies, and institutional and organizational development to support the bargaining position of rural farmers. CRA requires investments urgently in the following:

1. Managing climate risks
2. Understanding and planning for adaptive measures
3. Exploring options to reduce gaseous emissions, especially GHGs, wherever possible.

Elements of CRA: The following four are the important ones:

1. Managing crops, farms, livestock, aquaculture and fisheries, to enable manage the resources better.
2. Provide better ecosystem services and landscape management to conserve ecosystem enabling enhancement of efficiency of use.
3. Enabling famers to face climate risks and undertake mitigatory measures.
4. Changes in the wider food system including demand–supply and enhancing value chain interventions that enhance the overall benefits of CRA.

Management of crops, livestock, farms, aquaculture and capture fisheries: Options for crops management include switching varieties, species, changing cropping calendars, and nutrient management such as micro-dosing, the nutrient buffer power concept implementation (Nair 2019a) etc. Options for livestock management include pasture quality improvement, changing herd management etc. In the fisheries sector, pond management, introduction of local species should be considered. Overall farm management options include production diversification, integrated livestock management, also integrated pest management.

Landscape or ecosystem management: CRA also encourages examining agricultural systems in the context of larger landscape management, in order to understand better the linkages between agricultural production and ecosystem services.

Services for farmers and land managers: In order to enhance the adaptive farmer capacity, vital information such as climate changes, as seasonal forecasts or early warning systems, advisory services that link climate information to agricultural operations, and also financial services, such as credit and crop insurance are provided.

Changes in the wider food system: Agricultural production not only focuses on adaptation and mitigatory activities, but, also, help support food security and livelihoods, primary and secondary processing retail and consumer activities which are essential elements enabling and incentivizing an environment needed for climate smart agriculture.

Strategies for Climate Resilient Agriculture

The impacts of adverse climate can be minimized by adopting some strategies for land management, improved water management, resource conservation, crop diversification, use of tolerant/resistant varieties of both crops and livestock adapted to diverse climates, better weather forecasting, and improved pest management.

Effective Resource Management

Efficient resource management is very critical to the success of climate-smart agriculture. This is because there is substantial loss of food throughout the food chain. Therefore, energy use efficiency has to be enhanced as nearly 38% of annual energy is utilized in the food chain.

Integrated Renewable Energy Technology in Farming Systems

In order to create a stable change in energy smart and proficient food system, many technologies are required to be adopted. Such technologies may include wind mills, solar panels, biogas production, ethanol extraction as a bio fuel, bio energy operated water pumps, renewable energy powered vehicles, solar wind electricity production etc.

Availability of Technical Knowledge of Farmers

Conventional technical knowledge of people improved and developed in the course of time could offer a veritable source of ideas and feasible options for adoption to meet climate related challenges. Indigenous technologies with the farmers may be validated and used in climate smart agriculture.

Resource Conserving Technologies

Resource Conserving technologies (RCT) are required to reduce production cost and saving resources. Zero Tillage, is a remarkable example of this. No till farming can remarkably improve/alleviate the adverse impact of GHGs (greenhouse gases) and enhance carbon sequestration in soil (Zingore 2010).

Genetic Modification of Crops

Though the genetic engineering (genetic modification) has been a controversial subject, for long, there are instances where it has proved beneficial. One remarkable example has been in vegetable production, which has shown remarkable improvement during the past quarter century. When it comes to cereals and cotton, the results were dubious, and, very often deleterious and controversial.

Land use management such as shifting production between livestock and crops, reducing use of chemical fertilizers and pesticides, adjusting time of sowing and harvesting etc. have had a very favorable effect on meeting climate challenges.

Cropping Season Variation

Altering the planting/sowing dates of crops can coincide the crop growth period with the wet period of the season, which can help them withstand extreme adverse climate. Changed cultivation systems include improving the better cultivars and enhancing the intensity of farming various crops. Farmers will have to manage the changes in different hydrological regimes by adopting changed crop seasons (Pathak et al. 2012).

The Relocation of Crops

Adverse ambient temperature may adversely affect the quality of some crops, such as aromatic rice and tea. Other areas with cooler climates within a country or region will suit these crops better. This will assure better crop quality. Such areas within a region or country must be identified to accommodate these crops which are highly susceptible to climate change. This will be of pecuniary advantage to the concerned farmer growing these crops.

Efficient Pest Management

Variations in both ambient temperature and precipitation affect pest presence, of both insects and fungal and bacterial pathogens. Integrated Pest Management (IPM) which involves both use of resistant/tolerant crop cultivars and other cultural and biological control measures need to be adopted.

Carbon Sequestration

The natural and deliberate processes by which CO_2 is removed from the atmosphere or diverted from emissions and stored in the ocean, terrestrial vegetation, (vegetation, soil and sediments) and geological formations is known as "carbon sequestration". This is now the most important sought out mode to enhance soil fertility.

Methods of Climate Resilient Agriculture (CRA)

The following ago-ecological principles are to be followed to ensure CRA.

1. Increasing biomass in soil.
2. Strengthening the resilience of agricultural systems through enhancement of biodiversity, by creating habitats for natural enemies of pests. These details are elaborately discussed in the most recent book of Nair (2023) on agricultural biodiversity.
3. Managing soil organic matter levels to enhance soil biological activity and providing a favorable condition for optimum plant growth.
4. Minimizing energy loss, water, nutrient and genetic resources by conservation and regeneration of soil and water resources and ago biodiversity (Nair 2023).
5. Species diversification.
6. Enhancement of biological interactions and synergies among the components of agro-biodiversity, thereby promoting key biological/ecological processes and services (Nair 2023).

Some of the potential technologies of CRA are detailed below:

1. Use of quality seeds and planting material of crop varieties well adapted to withstand climate change: Quality seeds and planting materials of well-adapted cop varieties/cultivars resistant to abiotic stressed like drought, soil salinity etc. Plant breeding efforts involve multi-locational field trials to develop crop varieties which are tolerant/resistant to climate centered adversities. The farmers also obtain seeds from multiple unregulated sources, which include saving seeds from their own harvests, purchasing seeds from local markets, and, exchanging seeds with family members and neighbors. Such informal systems predominate in developing countries, such as India, where crop production systems are the most vulnerable to extreme weather events. Such informal seed supply has also to be community-based seed production and distribution channels. But, quality assurance is very critical in such endeavors as admixture with spurious materials will damage farmer confidence.
2. Biodiversity management: Crop diversity is especially important in a climate-smart approach to farming because it contributes best to pest and disease management, in addition to, meeting the challenges posed by climate change. Enhancing

on-farm biodiversity also provides other environmental services, including pollination, that are essential to farmers, as a whole. In a cropping system, greater biodiversity of crops and other living organisms is an important component of ensuring farm resilience, economic stability and profitability. Diversification is especially important in a climate-smart agriculture system as it contributes to ecologically sound pest management. Growing multi-purpose crop varieties can provide food, bio-fuel feed, fiber etc., to serve the multiple needs of farmers and consumers, besides ensuring climate resilience. The options include mixed farming, relay cropping, inter-cropping, crop rotation, cover crops etc. All these aspects are discussed in great detail in the recent book of this author titled "Biodiversity in Agriculture Sustainability of Soil, Soil Fauna and Soil Flora" published by Springer Nature International (Nair 2023).

3. Integrated Pest Management (IPM): Crops are usually adversely affected by various insects, soil pathogens, weeds and other harmful organism in a cropping system. At the same time, there are many beneficial organisms which provide a natural check on the harmful organisms. Indiscriminate use of chemicals and synthetic fertilizers have upset the natural balance of the ecosystem which has invited more of pest attack. The exclusive reliance on toxic chemicals for pest control will upset the pest-defender balance and result in pest resurgence. IPM addresses this question. It is a broad-based ecological approach which utilizes all the available tools of pest management judiciously to manage pest population within the economic injury level. The host-plant resistance obtained by the employment of resistant/tolerant crop varieties/cultivars, cultural control, physical and mechanical control, biological control and need based chemical control at a stage when the pest population is at a maximum level overshooting the economic threshold level (ETL) are used in the IPM approach. Farmer Field schools, which facilitate learning by doing and experimentation of different options by farmers, educate and train farmers to adopt IPM technologies. As climate changes, national regulatory, policy and institutional frameworks must be strengthened to enable the adoption of IPM on farms and in rural communities engaged in farming.
4. Improved water use management: Taking into account the most likely water use scarcity for agriculture, globally, it is imperative that enhancing water use efficiency in agricultural production is a dire need. Harvesting rain water can be accomplished by digging up ponds, constructing water harvesting structures and check dams for conservation of rain water. Water management (rain water and irrigation water) is very crucial in regions prone to climate risk. Where water is a limiting factor, improvement of water management can be achieved through measures which cover both soil and water, and/or with deficit irrigation that can maximize crop yields per volume of water applied. The principle should be "more crop per drop". In this context, it is important to note that both effective soil management and water management should go hand in hand, not exclusive of each other. It is important to remember that it is the electrically charged clay particle, with a net negative charge, which helps bond water—a net negatively charged clay particle attracting the hydrogen which is positively charged in a water mole, thereby building a chain of water molecule. Hence, all measures

of water management must be complimentary to good soil management, not exclusive of good soil management.

5. Sustainable soil and land management for enhanced crop production: Conservation agriculture is an approach which combines limiting soil Disturbance to the minimum, maintaining soil cover and diversifying crop production. Conservation agriculture provides a strategic entry point for climate change adaptation. Keeping the soil covered by cover crops reduces moisture loss, stabilizes soil temperature, reduces both water and wind erosion, restores soil carbon through the decomposition organic matter (crop residues especially) and provides food for beneficial soil microorganisms. Crop rotation, cover cropping, minimum tillage, organic recycling in soil by incorporating crop residues, mulching, addition of organic matter such as farm yard manure (FYM).
6. Sustainable mechanization: Efficiency enhancement in the use of farm machinery results in the enhancement of crop production. Using smaller tractors, making fewer passes across the field and reducing working hours, can reduce the emission of GHGs like carbon di oxide, and, minimize soil disturbance. This checks soil degradation and controls soil erosion.
7. Planting shade trees: Planting shade trees, protects the farm from the scorching sun and harsh winds and strong showers. Trees Planted as shelter belts can resist wind erosion.

Climate Resilient Action Plan

1. Climate resilient cropping patterns

There are several options for climate change adaptation and mitigation across different cropping systems depending upon types of climate stress, the farmers ability to cope and adapt, and the degree to which each climate factor affects crop yields. It calls for farming practices which uses, in the first place, quality seeds, and planting materials o well adapted materials; Cultivation of a range of crop species—varieties and cultivars—intercrops, IPM etc.

2. Climate resilient forestry

Apart from providing livelihoods for the rural communities by supplying food, fuel and fodder, medicinal herbs and other raw materials for industrial needs, forests act as buffers against climate shocks and provide ecosystem services which influence agricultural production. The forests protect soil and water resources, enhance soil fertility by shedding leaves and building up soil carbon stock, provide habitats for wild pollinators and predators of agricultural pets. Most importantly, they sequester soil carbon. And, of late, forests are a great interest in "carbon credits" scheme.

3. Climate resilient livestock production

Livestock is a very important component of farming system. In mixed farming systems, of crops and livestock, livestock consume crop residues and by product and the manure is used as a source of plant nutrients for the crops sown. The contribution of livestock for rural livelihoods goes far beyond agricultural production and food security, because it directly supports social welfare, education and health. Careful livestock management is required to maximize the range of services which they provide and reduce the adverse climate impacts. There are many synergistic options for both adaptation and mitigation which can be implemented in the livestock sector, including shifts in species and breeds, improved feed management, sustainable grazing practices, silvo-pastoralism and, on and off-farm diversification. Agro-silvo-pastoralism is a production activity that combines pastoralism (extensive livestock husbandry on pastures) and agriculture in a partially wooded environment. This activity closely combines these three modes of land use, and might even articulate them on a single site.

4. Climate resilient fishery

Inland fishery and captured fishery are adversely affected by climate change. Human population growth has escalated the demand for fish and fish products and unsustainable fishing practices have demanded more production from marine fisheries. Aquaculture will have to bridge the gap between shrinking supply and escalating demand. There are numerous options for supporting such growth in a climate-smart and sustainable manner, including sustainable water management practices, selective breeding and genetic improvements improved feed management, strengthened emergency procedures ad biosecurity measures.

5. Integrated production system

Integrated production systems are also called integrated farming systems where the by-product of one component is used as an input in the system. Examples of integrated production systems include agroforestry and crop-livestock combination, rice-fish combination, fish-livestock combination and food-energy systems. The diversification of resources and incomes associated with integrated production systems offers producers a greater number of risk management strategies and options to adapt to the adverse impacts of climate change. Also, the greenhouse gas emission intensities of integrated systems are typically lower than those of specialized systems. The sustainable intensification of production on integrated agricultural systems requires better understanding of the adverse impacts of climate change and climate variability on these systems.

6. Water management and climate change

Climate change affects the water cycle. Agriculture is affected by increased demand due to ambient temperature rise. The soil–water relationship is very dynamic, because, it is soil particles which hold the water molecules. An excessively negative charged clay particle attracts a positively charged hydrogen atom from the water molecule and thus a chain of water is held in soil. Hence, any change in the ambient temperature will have both a direct effect on soil and water. Changes in the amount and pattern of precipitation will affect sources of water for irrigation and also groundwater recharge. Globally, agriculture uses 70% of freshwater withdrawal for crop production. Rainwater harvesting, use of marginal water and wastewater resources, water use efficiency measures and watershed management procedures are some of the examples of options identified for effective water management to address the adverse impacts of climate change.

7. Sustainable soil and land management

In cropping, grazing, and forest systems, climate change will adversely affect soil health and plant growth. Erratic precipitation and continued drought will minimize soil capacity to supply water and plant nutrients. Higher transpiration and evaporation leads to erosion, reducing ground water recharge leading to reduced moisture for optimum plant growth, leading to soil salinization. GHGs emission can be controlled by proper land use management. Sustainable soil and land management initiatives which build up soil organic matter can provide a soil system better suited to act as a good and dependable ecosystem service, offering better climate resilience.

8. Genetic resources for food and agriculture

Selective breeding and domestication, combined with natural selection, have led to the development of numerous and diverse crop varieties, cultivars, breeds, stocks and strains of plants, animals and microorganisms. Such genetic diversity is lost following agricultural intensification through monocropping and unbridled use of synthetic fertilizers, and chemicals etc., such as herbicides and pesticides, through such highly chemical-centric and extractive farming as green revolution (Nair 2019c, 2023). As different plants, animals and microorganisms have different capacities to survive and adapt to shocks and changes, such as aberrant abiotic stresses like climate change, safeguarding and sustainably managing the diversity of genetic resources is critical to withstand severe and adverse impacts of climate change, such as the one observed globally in 2024 (Nair 2023). Characters such as capability to tolerate/resist abiotic stresses like tolerance to water stress, drought resistance, or biotic stresses like resistance/tolerance to pest and disease attack may be included in breeding or conservation strategies, designed to build resilience to climate change. Additionally, the indigenous, especially crop wild relatives, should also be preserved as these are better adapted to climate change and variability. These aspects have been, in particular, and, in great detail, discussed by this author in a chapter titled "Utilizing Crop Wild Relatives to Combat Global Warming" in Advances in Agronomy (Nair 2019c).

9. Energy smart food

Food production and post-harvest activities are energy-intensive. This has resulted in the emission of GHGs also. The agri food sector is estimated to contribute about 30% of total global GHGs emission. A sustainable approach must focus on water-energy-food nexus, and, address trade off and capitalize on synergies in the use of these resources. Food loses occur at all stages from production to post harvest activities, and this amounts to around 38% of the energy consumed by the whole food chain.

This could be compensated through the sale of biogas produced on farms to local households, or through the generation of electricity from agricultural residues which can be fed into the natural energy grid. To make the food chain more environmentally sustainable and climate-smart, it is important to examine the carbon foot print of the food chain, and, where possible, identify more efficient, less carbon-intensive approaches.

References

FAO 2011. Save and grow: a policy maker's guide to the sustainable intensification of small holder crop production. FAO, Rome

FAO 2012. The Sahel crisis: executive brief. March FAO, Rome. Fisheries

IFRI 2009. Climate Change: Impact on Agriculture and Costs of Adaptation. International Food Policy Research Institute, Washington, USA

Nair, K.P.P. 2019a. Intelligent Soil management for Sustainable Agriculture: The Nutrient Buffer Power Concept. Springer Nature Switzerland AG 2019

Nair, K.P.P. 2019b. Combating Global Warming: The Role of Crop Wild Relatives for Food Security. Springer Nature Switzerland AG 2019

Nair, K.P.P. 2019c. Utilizing Crop Wild Relatives to Combat Global warming. Advances in Agronomy, Vol 153, 175–258

Nair, K.P.P. 2023. Biodiversity in Agriculture Sustainability of Soil, Soil Fauna and Soil Flora. Springer Nature Switzerland AG 2023

Pathak, H., Aggarwal, P.K., and Singh, S.D. Editors. 2012. Climate Change Impact, Adaptation and Mitigation in Agriculture. Methodology for Assessment and Applications. Indian Agricultural Research Institute, New Delhi, India. pp. 302

Zingore, S. 2010. SOC sequestration in farming systems in Africa. Potential opportunities and challenges. In: 2nd Meeting of the Round Table on Organic Agriculture and Climate Change. RTOACC, 2010

Chapter 7
Which Are the Broad Strategies for Climate Resilient Agriculture?

Why the Need for Climate Resilient Agriculture Strategy?

Climate change has long lasting and potentially very lasting damaging consequences on the environment, composed of humanity, animals and plants, and, other soil biota. In short, it will gravely affect the agriculture sector, consequently food security. Adoption of climate smart agriculture has thus become inevitable in order to build resilience in the agriculture systems which will help cope up with climate change. With a view to implementing climate smart agricultural practices to ensure climate resilience, certain broad strategies have to be adopted. The CRA approach is designed to identify and operationalize sustainable agricultural development within the explicit parameters of climate change (Climate Smart Agriculture Source Book, FAO 2013). It integrates the three dimensions of sustainable development (economic, social and environmental), by jointly addressing global food security and climate challenges.

Different strategies for CRA are genera plans of action intended to address the adverse impacts of climate change in the long run. While preparing the strategies, the impacts of climate change on agriculture are kept in mind.

What Are the Climate Resilient Strategies?

1. There are three broad strategies for climate smart agriculture, as follows:
   a. A set of strategies adopted to increase crop yield under weather variability.
   b. A set of strategies adopted to make the agriculture system adapted to climate change.
   c. Mitigation strategies adopted to reduce the GHG (green house gas) emissions and managing the risk of climate change.

K. P. Nair, *Climate Resilient Sustainable Agriculture*,
https://doi.org/10.1007/978-3-031-84166-8_7

The CRA options integrate innovative traditional technologies, practices and services which are relevant for a specific location and reduce the effect of climate change. Substantial reduction of GHG gases emission over the years can reduce the occurrence of climate variability, increase the prospects for effective adaptation, reduce the costs and challenges of mitigation and contribute to eliminate climate-resilient pathways for sustainable agriculture.

d. Management strategies to increase crop yields under variable weather conditions

Agricultural technologies and practices, such as, minimum tillage, various methods of crop establishment, nutrient and irrigation and residue management, can improve crop yield, nutrient and water use efficiency, and reduce GHGs emission from agricultural activities. The two important farm management strategies which employ CSA methods to boost agricultural productivity are listed below:

(i) Sustainable soil management

Of all the soil management practices that address the CSA concerns, the most crucial is the nutrient factor, because, the nutrient factor decides more than 50% of crop productivity. In this connection a mention must be made of "**The Nutrient Buffer Power Concept**", a revolutionary fertilizer management technique that this author developed, working over three decades in Europe, Africa and Asia. The details are explained in the blockbuster book that this author has authored titled "Intelligent Soil Management for Sustainable Agriculture: The Nutrient Buffer Power Concept", published by world's number one science publisher Springer Nature (Nair 2019a). The author has employed thermodynamic principles in precisely assessing bioavailability of plant nutrients from which precise and economically feasible and cost effective fertilizer recommendations are made. The method has been adopted as a benchmark for the multimillion dollar USAID (United States Aid for International Development) funded soil testing program for West Africa, coordinated by Kansas State University's soil science department.

(ii) Preventing and mitigating soil degradation

The highly chemical centric farming, euphemistically called the "green revolution" has resulted in huge soil degradation in South Asia, in particular, in India. Of India's 328.73 million hectares of geographical area, more than 120.40 million hectares have now degraded soils, thanks to the green revolution. And this huge soil-related hazard coincided with the green revolution. In India, Punjab, known as the "cradle of green revolution" has thousands of hectares where not even a blade of grass will grow due to sol degradation. Scientific soil management, at the core of which is precise nutrient management, is the answer to this devastation.

(iii) Controlling soil erosion

Soil erosion is caused by weather and wind, water and anthropogenic activities. Investigations have revealed tillage-induced soil erosion as the major cause of severe soil carbon depletion and soil translocation on upland landscapes (Lobb et al. 1995).

(iv) Improving water storage

Many factors decide water storage capacity in soil, such as annual precipitation, soil depth, pore space, sol texture and soil structure. Soil management can influence rainfall water infiltration and the capacity of soil to reduce surface soil water evaporation and help store water in soil. Groundcover management, with the presence of huge amounts of well decomposed organic matter can improve soil water storage.

(v) Improving soil structure with organic matter

Usually, soils are compacted by repetitive plowing or hoeing and trafficking machinery (tractors) aggravate soil compaction. Enhancing soil organic matter content is the best solution to this problem. Specific cover crops with strong and deep roots, such as radish or pigeon pea (*Cajanus cajan*) can be employed to break up soil clods and ameliorate compacted soils.

(vi) Managing the organic matter for carbon sequestration

Elimination of perennial groundcover, repetitive cultivation, or continuous grazing, bare fallows, removal of crop residues and grassland burning (especially after tice/ wheat harvest, as practiced in Punjab State in India), etc., cause soil organic matter loss and depletion of the water holding capacity of soil. Monoculture of cash crops and the unbridled use of external inputs, such as synthetic fertilizers, also leads to soil carbon loss. Production of energy intensive synthetic fertilizers and pesticides and use of fossil fuel in farm machinery are major sources of GHG gases emission. Further when incorrectly applied, these inputs leach into ground water leading to water pollution, making water non potable, and ultimately become deleterious to human health.

Diversified cropping, crop rotations and improved techniques for fertilizer management (Nair 2019a) help cut down on GHGs emission.

(vii) Water management for sustainable agriculture

Climate change accelerates crop evapotranspiration, changing rainfall patterns, leading to greater variations in groundwater recharge, including weather aberrations. The adverse effects of unprecedented droughts and floods leading to groundwater depletion severely affect water management. Sound water management technologies and automated irrigation systems can help accelerate better irrigation and ensure substantial water saving.

(viii) Adaptation at farm level

Farmers prefer more efficient irrigation technologies which cut down to surface evaporative losses. These technologies can be combined with deficit irrigation technologies to maximize production per volume of water consumed. Crop selection and changes in crop calendars will help farmers adapt to altered temperatures and rainfall patterns stemming climate change.

(ix) Adaptation at irrigation level

Modern irrigation systems, automated irrigation and using water based on tensiometer reading etc., can economize irrigation water use and enhance water use efficiency. Intermediate storage of water and conjunctive use of surface and ground water sources can economize use of irrigation water and build resilience and reliability of water supply.

(x) Adaptation at watershed, river basin, and national levels

Flood management plans requires infrastructure upgradation with non structural, information-rich approaches tailored to mitigate the flood impacts better. Similarly, there will be a need to shift from drought to emergency response to drought management plans. Improved weather forecasting and hydrological monitoring will become a critical element of modern adaptation strategies.

(xi) Water management for climate change mitigation

Normally irrigated agriculture uses agro chemicals for higher yields and ground water exploitation is heavy. Groundwater use is expanding both in absolute and relative terms and this increases fossil fuel consumption, hence energy costs. In India, it is responsible for 4–6% of the national total emissions due to this activity. Agricultural methane (CH_4) emissions account for more than 50% of methane emissions from anthropogenic (human) activities. Flooded rice cultivation releases methane. In this connection SRI (System of Rice Intensification) is a water conserving measure, where, younger seedlings are transplanted in a square pattern under puddled conditions in the field without submergence to allow more number of tillers to emerge. Weds are incorporated into the soil. Since SRI reduces the amount of flooding of irrigated rice, it reduces the methane emissions and conserves irrigation water. It may also reduce nitrous oxide emissions (HLPE 2012).

Adaptation Strategies

Adaptation strategy is very important to reduce the severity of climate change impacts on agriculture. Different approaches can be taken up to adopt these strategies.

Adaptation Through Transgenic Techniques

Transgenic approach, or, what is known as genetic engineering, helps transfer, desirable genes through the RNA-recombinant technique. Resistance to drought, salinity etc. can be achieved through this technique. This approach has also been proposed by this author with reference to crop wild relatives in his book titled "Combating Global warming: The Role of crop Wild Relatives for Food Security" published by Springer Nature (Nair 2019b).

Heat Resistance

Plant have various mechanisms to cope up with high ambient/soil temperatures. Such mechanisms refer to maintaining cell membrane stability, or buy ion transporter, proteins, osmoprotectants, anti oxidants etc. Crop cultivars with such characteristics tolerate/resist adverse climate changes.

Drought Tolerance

Several drought tolerant transgenic plants have been developed by plant breeders through genetic engineering. Some important proteins like protein kinases, receptor-like kinases, transcription factors, some enzymes related osmoprotectant characteristics or plant hormone synthesis, and, other functional regulatory proteins play an important role in defence against drought stress.

Salinity Tolerance

Climate change leads to accumulation of salts in plant cells. These salts, especially of sodium, lead to salinity. Transgenic plants accumulate more potassium, calcium and magnesium ions and much less sodium ions.

Developing Climate Ready Crops

Several drought tolerant crop varieties have been released for farmers' cultivation in India. Some notable ones are, *Sahbhagi dhan*, IR64-Drought-1 among rice crop. Singh et al. (2014) have reported 10% higher yield with *Sahbhagi dhan.*

Crop Diversification

Instead of monoculture, intercropping, relay cropping and crop rotation have been found useful to combat climate change. Intercropping of soybean and pigeon pea in the ratio of 4:2, pearl millet and pigeon pea in the ratio of 3:3, pigeon pea and green gram in the ratio of 1:2 and cotton and green gram in the ratio of 1:1 have been found more economic than mono cropping (Prasad et al. 2014). Nitrogen fixing pulse crops have a lower carbon and water footprint as compared to other crops.

Alteration in Land Use Pattern

Change in location of crop and livestock, adjustment in cropping patter, planting time and methods of fertilizer application and pesticide use pattern etc., will address climate change. Alternate use such as tree farming. Silvipastoral system, agroforestry are also recommended.

Adjustment of Planting Time

Both sowing and harvest time can be manipulated to variable weather conditions.

Conservation Agriculture

Conservation agriculture (CA) enhances natural biological processes above and below the soil. It is characterized by following there inter-lined principles:

1. Minimum mechanical soil disturbance
2. Permanent soil cover by mulching
3. Diversification of crop species grown in sequence/association for risk minimization. Conservation agriculture-based resource conservation technologies (RCT) under precision laser land leveling are helpful to improve water productivity.

Increasing Resource Use Efficiency

Resource conserving practices include minimum tillage/zero tillage, use of neem-coated urea for maximizing nitrogen use efficiency, crop residue management, irrigation management etc. RCT technologies help conserve natural resources like soil, water and biodiversity.

Integrated Nutrient Management (INM)

INAM practice integrates application of plant nutrients though organic and inorganic sources for soil health and productivity. Based on this, it can be recommended that substitution of 50% of recommended N with organic sources increases crop yields (Prasad et al 2016). INM can reduce the production cost and provide climate resilience. There is now great focus on organic sources of fertilizers. In this connection a revolutionary soil management technique with specific reference to fertilizer management can be found in the book authored by this author (Nair 2019a) titled "Intelligent Soil management for Sustainable Agriculture: The Nutrient Buffer Power Concept" published by Springer Nature.

Site-Specific Nutrient Management (SSNM)

SSNM provides principal tools for nutrient supply as and when the crop needs to achieve high productivity using indigenous source. Nutrients should be applied at the right time, right quantity and right methods. Leaf Color Charts (LCC) and GreenSeeker are accessed to determine precise nutrient application times.

Relocation of Crops

Climatic variability such as increased ambient temperature, CO_2 level, drought and floods would adversely affect crop performance. There are sensitive crops which have to be identified and relocated.

Harnessing Indigenous Technical Knowledge of Farmers

Indigenous technologies (ITKs) are available in rural areas which provide adaptation to climate variability. These technologies may be validated and fine-tuned o overcome climate variabilities. Such ecological-based ITKs can be used as adaptive measures.

Integrated Farming System (IFS)

For farmers, monocropping is a risk filled enterprise for farmers who farm in flood prone and drought prone areas. IFS is defined as the integration of different inter-related, interacting and interdependent farm enterprises which are suited to agro-climatic conditions, socioeconomic situations of the farmers. IFS minimizes financial risks arising from natural calamities and generates higher income. An example is integrated fish-duck farming and rice-fish-poultry farming for the benefit of small and marginal farmers (Prasad et al. 2014).

Integrated Pest Management (IPM)

The dynamics of pest population and crop diseases is rapidly changing owing to extreme climate stress. Many climatic factors, such as, variability in precipitation, changes in ambient temperature, also soil temperature, consequent to the steep increase in ambient temperature, will have a severe effect on insect pest and diseases incidence, also pest/weed population relationship, ultimately, adversely affecting pest-weed-host interaction. Under ideal field conditions, crops pests are controlled by natural defenders. The farmers very often use agro chemicals, indiscriminately, to control crop pests, without realizing the important role of natural defenders, which becomes counterproductive operation. Indiscriminate use of agro chemicals, most of which are toxic by their very nature, leads to resurgence of crop pests, without the farmers realizing that these pests develop resistance to such agro chemicals. IPM provides a lasting solution to pest management. It uses resistant/tolerant crop varieties or crop cultivars, mechanical control, biological and need-based chemical control, when the pest population overshoots the economic threshold level (ETL). Many effective bio-agents are now available in the market also from laboratories which conduct research on this theme, and there are many improved traps, such as pheromone trap, yellow trap and sticky trap along with bio-pesticides.

Early Warning System

This is based on weather foresting. Farmers are advised based on weather pattern how best to go about their field operations. Information and Communication Technologies (ICT), can greatly help the administrators And researchers in developing contingency crop plans which also reduces the risk of aberrant weather. Basing the local weather data, early warning system provides farmers the directions to take their crop cultivation activities in the right direction, along with crop advisory, to reduce the adverse impact of climate stress.

Weather-Based Crop Insurance

Loss in crop production, consequently farm income, emanating from unforeseen weather aberrations, is beyond the carrying capacity of poor and marginal farmers. The State and National governments are coming up with relief measures, such as, suspension of land revenue taxes, farmers loan waiver, calamity relief fund etc., in India. In India, since 1972, agricultural insurance schemes have been in operation, which has been a great boon to farmers. In recent years, weather index insurance contracts in agriculture have emerged as a better alternative to the traditional agricultural (crop) insurance schemes.

Mitigation Strategies

Mitigation practices helps reduce the adverse impacts of gaseous emission hazards such as GHGs emission. A change in the system of cultivation, cultural practices, nutrient management practices, also water management practices, help rescue GHGs emission. In this connection, the importance of "The Nutrient Buffer Power Concept" on sustainable agriculture discusses, a revolutionary idea on fertilizer use management is very relevant (Nair 2019a).

The following paragraphs discuss some of the important green house gases.

Methane

Methane is about 25 times more effective than carbon dioxide in its heat trapping efficacy. In agricultural production, methane gas is released from wetland rice cultivation, organic matter decay, enteric fermentation of cattle dung etc. Production, oxidation and transport of methane gas from paddy fields can be done by alteration of water management in paddy fields, in particular, by promoting mid-season aeration, by short-term drainage and direct seeded rice (DSR) cultivation. Reduction in methane emission increased further when DSR was combined with mid season drainage or intermittent irrigation, compared to flooded transplanted rice. The reason for low methane emission from DSR is aerobic conditions in the soil, especially during the early growth stage of the rice crop. The other methods for reducing methane emission are improved organic matter management by promoting aerobic degradation through incorporating or composting it into the soil during off-season soil drained period, cultivation of rice varieties with high root oxidative activity and high harvest index, and feed management of the cattle by employing improved feed additives, substituting low digestible feed with high digestible ones, and grazing, which can reduce emission of methane in livestock farming.

Carbon Dioxide

The principal sources of carbon dioxide emissions are decay of organic matter, forest fires, eruption of volcanoes, burning of fossil fuels, deforestation and land-use changes. Such emissions can be controlled by carbon sequestration, efficient sol management practices, and planting trees. Carbon sequestration is one of the most efficient strategies to mitigate CO_2 emissions, which is, in practice, removal of atmospheric CO_2 and converting it into soil carbon on a long-term basis.

Nitrous Oxide

Nitrous oxide is a lot more effective in heat trapping property compared to carbon dioxide. Nitrous oxide enhances the greenhouse effect just as carbon dioxide does by capturing reradiated infrared radiation from the Earth's surface and subsequently warming the troposphere (lower atmosphere). One ton of nitrous oxide traps 273 times more heat than one ton of carbon dioxide and can persist in the atmosphere for up to 110 years, compared to 10 years for methane, according to the National Aeronautics and Space Administration (NASA), USA. The highly chemical-centric green revolution also contributes to the emission of nitrous oxide, through the microbial decomposition of unbridled use of urea to prop up crop yields (Nair 2019a). According to this author, nitrous oxide can contribute up to about 30–35% to global warming. Site Specific Nutrient Management (SSNM) in crop fields, integrated nutrient management (INM), use of slow release nitrogenous fertilizers, nitrification inhibitors, use of neem-coated urea etc., can reduce N_2O emission.

Carbon Sequestration

Trees, plants and crops absorb CO_2 naturally through photosynthesis and it is stored as carbon in biomass, in tree trunks, branches, foliage and roots and soils. Trees function as carbon sinks and store large amounts of carbon in their vegetation and root system.

Soil Management Practices

Soil management practices, such as reduced or no tillage, residue incorporation and mulching can improve soil biodiversity, micro-aggregation and improve soil health and water holding capacity of soil. Conservation tillage reduces soil disturbance, due to mechanical interventions, and help mitigate the release of soil carbon to the atmosphere. Some of the important management questions concerning soil, especially with regard to fertilizer applications, are discussed by this author in his book titled Soil Management for Sustainable Agriculture "The Nutrient Buffer Power Concept" (Nair 2019a).

Improved Cropping and Organic Matter Systems

Organic systems of crop production increases sol organic matter levels through the use of composted animal manure and cover crops. Addition of sufficient organic manure can dispense with the application of synthetic chemical fertilizers. Reduced carbon dioxide emissions and increased soil carbon stock can be achieved by direct seeding, field windbreaks, rotational gazing, perennial forage crops, reduced summer fallow and proper straw/residue management. Using higher yielding varieties/cultivars to maximize crop yields, can also increase soil carbon sock. Conservation tillage, integrated and site-specific nutrient management, crop rotation and diversification, efficient management of animal manure and feed management in livestock can reduce GHGs emission.

Land Restoration and Land Use Changes

Land restoration and land use changes which encourage conservation and soil improvement, air quality etc., can reduce GHGs emission.

Livestock Management

About one-fifth of global methane (CH_4) emissions originate from enteric fermentation in cattle. Proper animal breeding with a focus in feed efficiency can reduce methane emissions. Changing animal feed, grazing, better forage use ca reduce methane emission reduction.

Mitigation Through Transgenic Crops

Transgenic crops can contribute to reduction of carbon emissions through reduced fossil fuel-based chemicals' use in their cultivation. Nitrogen fixation is also enhanced by some transgenic crops. However, the use of transgenic crops in agriculture has been a controversial subject.

Irrigation and Water Management

Improvement in water use efficiency coupled with a reduction in operating hours, drip and trickle irrigation, can significantly reduce the amount of water used for agricultural purposes, so also the amount of nitrogen fertilizer applied to the crop.

Nitrogen Use Efficiency

Precision farming and use of site-specific nutrient application practices can significantly reduce the emission of GHG. Other strategies include the use of cover crops and manures (both green manures like Sunhemp (*Crotalaria juncea*), *Gliricidia miaculata, Sesbania* etc., and Farm Yard Manure–cattle manure composted with other organic material like post harvest stubbles), crop rotation with n-fixing crops like beans, application of compost and bio fertilizers with better fertilizer application methods required to enhance nitrogen use efficiency.

Methane Capture

Large emissions of methane and nitrous oxide are attributable to livestock-waste treatment, especially in dairies. Agriculture methane collection and combustion systems include covered lagoons and complete mix and plug flow digesters. Bio gas plants have become very popular in India in many areas for methane production and methane gas is used for lighting, cooking etc. Additional benefits include reducing odors from livestock manure and reducing labour costs associated with removal of cattle manure.

Biofuels

Biofuels in the form of bio diesel or ethanol are carbon neutral. Biofuels can be produced from crop residues and sugarcane trash. Renewable energy sources, such as, wind and solar energy, also, present significant opportunities for the agriculture sector in achieving reduction in GHG gases' emission. Brazil is one country in the world which has brought about an "ethanol revolution" in the production of this renewable energy, primarily from sugarcane trash, which is so very commonly used in the place of petroleum products for locomotion.

References

Lobb, D., Kachanoski, R.G., and Miller, M.H. 1995.Tillage translocation and tilage erosion on shoulder slope landscape positions measured using 137Cs as a tracer. Canadian Journal of Soil Sci. 75, 211–212

High level Panel of Experts on Food Security and Nutrition (HLPE) 2012. Climate change and food security. A report by the HLPE of the Committee on World Food Security and Nutrition, FAO, Rome

Nair, K.P.P. 2019a. Intelligent Soil Management for Sustainable Agriculture: The Nutrient Buffer Power Concept. Springer Nature Switzerland AG 2019

Nair, K.P.P. 2019b. Combating Global Warming: The Role of Crop Wild Relatives for Food Security. Springer Nature Switzerland AG 2019

Prasad Y.G., Maheswari, M., Dixit, S., Srinivasa Rao, C.H., Sikka, A.K., Vekateswaralu, B., Sudhakar, N., Prabhu Kumar, S., Singh, A.K., Gogoi, A.K., Singh, Y.V., and Mishra, A. 2014. Smart Practices and Technologies for Climate Resilient Agriculture. Hyderabad, India. Central Research Institute for Dryland Agriculture (ICAR). p. 76

Prasad, J.V.N.S., Rao, C.S., Srinivas, K., Jyothi, C.N., Venkateswaralu, B., Ramachandrappa, B.K., Dahanapal, G.N., Ravichandra, K., and Misha, P.K. 2016. Effect of ten years of reduced tillage and recycling of organic matter on crop yields, soil organic carbon and its fractions in Alfisols of semi arid tropics of southern India. Soil & Tillage Research 156, 131–139

Singh, A., Yadav, A., Mathew, K., Varrier, V., and Singh, S. 2014. Success of IRRI stress tolerant rice varieties in Bihar, India-Experience of IRRAS farmers. In: Poster Presented at the Fourth International Rice Conference, 27 October–01 November 2014. Bangkok, Thailand

Chapter 8
Climate Resilient Agricultural Practices

Climate Resilient Agricultural Practices and Indicators

Considering the three pillars of CRA, such as, productivity, adaptation and mitigation, different technologies and practices have been evolved. The CRA practices are grouped under soil and land management, water management, crop management, livestock management, forestry, fishery and aquaculture, and energy management. The World Bank Group has suggested broad indicators of CRA, which were given in the earlier chapter.

Soil and Land Management

These practices can adapt to the adverse impacts of climate change and minimize the green house gas emission's adverse effects. They are detailed below:

Preventing Land Conversion and Protecting Vulnerable Lands from Loss of Organic Matter

Cultivated lands vulnerable to loss of soil organic matter are at a greater risk which needs protection. Soil carbon stocks in semi arid environments can decrease by as much as 30% in less than five years when native vegetation or pastures are converted to cropland (Noellemeyer et al. 2008). Cultivation of tropical forest soils causes losses of more than 60% of original soil organic matter (SOM) content stocks in a few years (Brown and Lugo 1990). Although drylands have lower mitigation potential

K. P. Nair, *Climate Resilient Sustainable Agriculture*,
https://doi.org/10.1007/978-3-031-84166-8_8

per hectare than humid soils, their overall contribution could be highly significant since dry and sub-humid lands cover 47% of the Earth's land surface. Moreover, many dryland regions affected by land degradation are sub-optimally managed which could rapidly respond to improvements in soil management (Farage et al. 2007).

Assessment of land resources is required to understand the extent of land degradation and identify hot spots in terms of land degradation. Suitable land use planning and policy support or incentives are required to reduce the impact of land degradation and reduce land conversion and promote the adoption of sustainable practices, with specific attention to lands which are more vulnerable to climate change impacts.

Preventing and Mitigation of Land Degradation

Degraded soils are vulnerable to serious loss of soil organic matter (SOM), and soil biodiversity, grater soil compaction, and increased rates of soil erosion and landslides. Land degradation is physical degradation by water ad wind, soil crusting, scaling and water logging; biological degradation due to depletion of SOM and loss of soil fauna and soil flora; chemical degradation due to soil acidification, nutrient depletion, soil pollution by unbridled use of synthetic chemical fertilizers and pesticides and herbicides, or through soil application of human waste and industrial waste, as well. The major reasons for soil degradation are continuous cropping, most of the time monocropping, without crop rotation and fallowing, repeated tillage, over grazing by cattle, burning of crop stubbles and nutrient mining, caused by "green revolution". Sustainable management practices which would reverse or minimize soil degradation, and restore soil biodiversity, sequester soil carbon, reduce or check GHG emission and ensure sustainable productivity would be urgently required.

Controlling Soil Erosion

Practices to reduce soil erodibility and surface run off are usually adopted for erosion control and soil conservation. Soil erosion caused by rain water, wind etc. can be prevented by several measures, as detailed below:

1. Contour faming
2. Terrace farming
3. No till farming
4. Organic faming
5. Restoring wetlands
6. Planting vegetative cover and forest restoration
7. Planting buffer strips along running stream banks
8. Windbreaks
9. Proper waste disposal management.

Minimum or no tillage can greatly control water erosion. Cross-slope vegetation, terrace cultivation on hilly terrain, contour bunding, tied ridges, grassy waterways, trenches, field bunding etc., can greatly control water erosion.

Improving Water Storage

Water storage in soil is governed by such factors, as, soil texture, soil structure, soil depth, total precipitation etc. Since surface evaporation, evapotranspiration from plants, and ground water movement influence loss of water from soil, soil management can influence rainwater infiltration and the capacity of the soil to reduce soil water evaporation and increase water storage in the soil. The soil management practices which are required to improve water storage in soil are minimum soil disturbance, which can be attained through no tillage or minimum tillage practices, mulching, precision farming, and micro irrigation like drip or sprinkler irrigation.

Improving Soil Structure with Organic Matter

Soil organisms break down the organic matter to use it as food for sustenance, and in tur release excess nutrients into the soil to be later utilized by plants for their sustenance. Soil organic matter primarily contains organic carbon (on average 58%, which, very much depends on the type of organic matter present in the soil), but also, essential nutrients for plant growth, and sometimes inorganic carbon, as well. It influences the physical, chemical and biological properties of soil. The factors responsible for soil organic matter decomposition are, the presence of the microorganisms, optimum soil temperature, optimum soil moisture content, and physical and chemical properties of the soil, in question, along with land use pattern and natural resource management practices.

Minimum Tillage

The primary concern of soil tillage is to provide a good seed bed and control weed growth. Tillage can be defined as the mechanical manipulation of soil with specific mechanical implements. Soil compaction is often caused by the use of heavy tillage implements in the field. In grazing lands, soil is compacted by the trampling of bullocks used to drag the implements. This can lead to restricted root growth. Hard pans develop. Tillage disturbance in soil causes reduction in carbon stabilization. Further, tillage of soil stimulates microbial decomposition of soil organic matter and this leads to the emission of emission of CO_2 into the atmosphere.

The management practices to prevent soil compaction and hard pan development, are minimum tillage or no tillage, plant or liter cover to provide organic matter which enhances the activity of soil fauna, for example, earthworms and termites. The burrowing of earthworms can break up compacted soil layers and incorporating SOM from the surface into the soil. Practices such as conservation agriculture, which combines minimum soil disturbance with increased soil cover and crop diversification will permit SOM to build up and the soil resilience to climate change.

Minimum tillage and conservation agriculture improve the biological, chemical and physical properties of the soil. Minimum tillage reduces time, labor and machine operations which help conserve soil moisture and reduce soil erosion. During the last few decades, advancement in weed control science have permitted many crops to be grown with minimum tillage. The following are the three types of reduced tillage:

1. Zero tillage (also called no till farming or direct seed drilling)
2. Conservation tillage
3. Minimum tillage.

Zero tillage helps build soil water reserves. In zero tillage crop residues are spread on soil surface, weeds controlled by cover cropping or non polluting desiccant herbicides as pre-pant application. Conservation tillage increases the soil's ability to store or sequester soil carbon. Minimum tillage helps increase soil water reserve and plant growth.

Managing Soil Organic Matter for Carbon Sequestration

Soil biota, temperature, both ambient and soil, soil moisture and chemical and physical composition, governs the organic matter decomposition in soil. Former land use and natural resource management, such as tillage, also govern it. Loss of soil organic carbon (SOC) and water holding capacity of soil is associated with practices such as elimination of perennial groundcover, repetitive cultivation, such as monocropping, continuous grazing, bare fallows, removal of crop residues and stubble burning. The monoculture of cash cops uses energy-intensive fertilizers and pesticides which cause GHG emissions. Excess and improper use of such chemicals leach into water resources and the resulting water contamination has serious deleterious effects on both ecosystems and human health. Crop rotations, and improved fertilizer use management, better seeds and pesticides can increase resource use efficiency and reduce GHG emissions. Improvement of soil structure and soil biodiversity, no till cultivation and the control of soil compaction will reduce GHG emission and enhance carbon sequestration in soil.

Nutrient Management

In conventional agriculture organic manures are supplemented by synthetic chemical fertilizers to supply to the crop both macro and micro nutrients. Excessive nutrient mining of the soil has to be properly managed by precise soil management. Fertilizer Tree Systems (FTS) involve the panting or regeneration of fast-growing nitrogen-fixing trees or woody shrubs to produce enough biomass and to adapt to local climate. Vermicompost is better source of organic manure than the commonly used farm yard manure (FYM). Many crop residues and animal manures are of low quality, as they fall below the critical nitrogen content of 1.8–2.0% and immobilize nitrogen in the soil temporarily, thereby exacerbating nutrient deficiency. A strategy used worldwide in intensive cropping systems is to combine inputs of organic matter (plant mulch, compost, crop residues and green manures, such as Sunhemp (*Crotalaria juncea*), *Gliricidia maculata* and *Dhaincha*). This will address or pre-empt both macro and micro nutrient deficiencies and help mitigate adverse climate impact.

Through the activity of soil microorganisms, organic nitrogen in soil is mineralized as ammonia, NH_3, and then into ammonium ions (NH_4) which is then oxidized to nitrites (NO_2) by *Nitrosomonas* and subsequently oxidized onto nitrates (NO_3) by *Nitrobacter*, which can be directly taken up by plant root. Since nitrate ions move rapidly in soil they will be directly taken up by plant root. There is also the possibility of its leaching out since it moves in soil very rapidly. This is particularly so in tropical soils. This is the reason why the nitrogen recovery from applied synthetic fertilizers is very low in tropical soils. Use of slow release nitrogenous fertilizers, nitrification inhibitors, and, even coating of the urea granules with neem seed cake, enhances nitrogen recovery.

There is common consensus that zero tillage and conservation agriculture systems reduce nitrate leaching considerably (Macdonald et al. 1989). Cover crops absorb the nitrogen from applied fertilizers and minimize leaching losses. *Biochar*, a stable and carbon-rich form of charcoal is produced by pyrolysis, where biomass is heated with little of no oxygen (Sparkes and Stoutjesdijk 2011). Possible biomass sources for *biochar* include milling residues, such as, rice husks, sugarcane bagasse, municipal wastes and animal manure. Soil application of biochar enhances crop growth due to its soil nutrient status enriching capacity.

Contour Farming and Terrace Farming

Contour faring is practiced in hilly regions along the contour lines with ridge and furrow system of planting to change the direction of the runoff. This is used in orchards, vineyards and nut crops farming. Contour farming minimizes sheet and rill erosion, also transport of sediments, other solids and contaminants attached to them. Bench terraces are formed along the contour and across slopes for planting on hill slopes.

Mulching and Residue Management

Mulching and residue management is a technique whereby at the time of the sown crop emergence, at least 50% of the soil surface is covered with organic residues from the previous crop. Mulching improves water infiltration, and the soil cover provides effective protection against rain splash erosion and surface runoff.

Sustainable Mechanization

Appropriate small machinery can carry out sustainable crop management practices which could increase production efficiency of farms. Small tractors used for small working hours combined with conservation agriculture can minimize soil disturbance, reduce carbon dioxide emission and reduce soil erosion and land degradation. These smaller tractors including two wheeler tractors use minimal fuel for locomotion. Equipment used in precision farming can enhance efficiency in the use of inputs and reduce direct or indirect GHG emissions. Investment in sustainable mechanization enables farmers to expand the range of their activities and diversify their livelihoods in ways that reduce their vulnerability to climate change. Battery-powered drones can eliminate GHG emissions. Solar pumps can help in irrigation.

Water Management

Climate change will impact every element in the water cycle. Agricultural production will be impacted by enhanced evaporative demands of crops, changes in the quantum of precipitation, and variations in the river runoff and ground water recharge which are the basic sources of irrigation. The hydrological cycle is expected to accelerate as rising ambient and soil temperature increases evaporation from land surface area (Turral et al. 2011). Following rise in temperature, evapotranspiration rate increases leading to depletion of soil moisture.

Option for adaptation to climate change will need to be applied at differ scales: field and farms, irrigation schemes, specially in large ones, watershed or aquifers, river basins (including trans-boundary river basins), and, at the national level.

Adaptation at Field and Farm Level

Farmers will be inclined to adopt more efficient irrigation technologies that minimize evaporative loses. Deficit irrigation, cop selection and changes in crop calendars, use of crop/varieties/cultivars with better resilience to climate change, especially dry spells, increased diversification in agricultural production, such as, incorporation agro-forestry etc. Farmers will also need to adopt relevant measures to respond to increased frequency of floods etc.

Adaptation at Irrigation Level

In the context of climate change, irrigation reduces evapotranspiration losses and saves crops from dry spells. Governments incur he expenditure in irrigation schemes to help farmers. In climate-smart agriculture, farmers are motivated to use sprinkler and drip irrigation which can save up to 70–90% in water use. Conjunctive use of surface and ground water along with rain water harvesting will augment water saving.

Irrigation modernization requires better water allocation mechanisms, the clear transmission of alerts about water scarcity to farmers, and the adaptation of both infrastructure and management for more flexible and reliable delivery of water. Intermediate pond storage of water, combining both ground and surface water, to reduce misuse of water etc., are some of the options. Management of water allocation, sensible cropping patterns and more effective water distribution by water users association (WUA) are simpler and more effective measures to induce productive water use at the irrigation system level.

In order to face climate change, irrigation would be required to counter the predicted increase in evapotranspiration, and, meet the contingency of reduced precipitation. However, drip or sprinkler irrigation is required. These systems are 70–90% more efficient in water saving. Photovoltaic or solar powered drip irrigation systems or solar powered pump can be useful in this manner. In existing irrigation systems conveyance loss should be minimized to enhance water use efficiency.

Adaptation at Watershed, River Basin and at National Levels

Meeting climate stress at higher levels require a combination of policy adjustments and investments in infrastructure and management. Adjustments in the storage capacity and management of dams and river protection works would be required at river basin level. Flood management plan will need to be based on approaches which can better mitigate the flood impacts through a combination of land planning, early warning systems and insurance schemes.

Table 8.1 In situ rainwater conservation practices

Soil type	Rainfall	In situ moisture conservation techniques
Red soil	Low	Dead furrow at 3–6 m interval
	Medium	Sowing on flat beds and ridging later
	High	Graded border strips
Black soil	Low	Contour cultivation
	Medium	Dead furrow at 3–6 m interval
	High	Graded open furrow at 10 m interval across the slope

Rain Water Harvesting

This involves collection and storage of rain water in surface or sub-surface aquifers, before the rainwater is lost through surface runoff. Rain water conservation practices can be divided into two main categories:

1. In situ management
2. Ex situ management.

In the former, rain water is harvested when it falls on land surface. In the latter, rain water is collected and stored for different productive use, such as, drinking purposes, such as, irrigation, sanitation purposes etc. Table 8.1 details the in situ conservation methods.

Location specific measures for in situ moisture conservation are conservation tillage, deep tillage, contour cultivation, dead furrows, strip cropping, contour farming, mulching, broad bed and furrow (BBF), vegetative barrier, ridge and furrow and tied ridging etc. These practices are taken up to protect the land from degradation and to encourage storage water for crop use.

In in situ rain water harvesting, different practices, such as, off-season tillage, tied ridges, mulching, retention crop residues, contour planting, ridging/furrowing, terracing, vegetative barriers, stone lines and planting basins etc. are practiced.

Water Management for Climate Change Mitigation

Since irrigated agriculture uses more of synthetic chemical fertilizers, and other agro-chemicals, than most rain-fed agriculture, efforts are needed to minimize GHG emissions. And this can be achieved through improved crop management practices. Ground water exploitation is increasing the use of fossil fuels and raising the costs of water supply. Renewable sources of energy use is an option to minimize GHG emission.

One-third of methane emission from agriculture comes from flooded soils in rice production. Emissions during growing season can be controlled/reduced by using various water management practices, such as, cultivating “aerobic” rice (without flooding), and, where conditions permit, adoption of alternate wetting and drying technique. System of Riced Intensification (SRI) reduces the amount of flooding of irrigated rice and, it is also likely to reduce methane (CH_4) emissions. It saves water and other inputs. Better pasture management, combined with the use of feed additives that suppress methane fermentation in ruminants, can reduce methane emissions substantially (Turral et al. 2011).

Water Management in Rice

Rice is a water guzzling plant and consumes maximum water for its good productivity. The alternate wetting and drying method implies that the rice fields are not kept continuously submerged or waterlogged, but, intermittently dried during its growth period the SRI method. The method is very relevant for climate change adaptation and mitigation. Flooded soils such as those for irrigated rice produce methane, a GHG (greenhouse gas) which plays a significant role in global climate change. The research done at IRRI and national rice research institutes in China, Indonesia, India, the Philippines and Thailand concluded that when rice fields are dried, oxygen becomes available in the root zone and this reduces methane emissions. Producing more rice with less water from irrigated systems could provide opportunities to improve human health as some diseases, such as, malaria, can spread under submerged conditions.

Climate-Smart/Resilient Crop Production

Several efficient crop management practices which address all the three aspects of climate smart/resilient agriculture should be considered as "climate smart/resilient" practices. The principles of climate smart/resilient crop management sustain crop yield by increasing input use efficiency and enhancing intrinsic system of climate resilience. The practices are, use of abiotic stress crop varieties, change in cropping pattern, adoption of farming system approach and propagation of ecosystem management options. Integrated soil-water crop management increases resilience to climate change.

Important crop management practices I the context of climate smart/resilient agriculture include direct seeding, in the case of rice production, in combination with protective soil cover, crop diversification and crop rotation, the elimination of the burning of crop residues, integrated soil fertility nutrient management to enhance the soil's nutrient retention capacity and the availability of nutrients to crop plants, the precise management of the most important nutrient, nitrogen, and integrated pest management.

It has been predicted that billions of people living in the developing countries will face shortages of water and food and greater risk to life due to climate change.

Although some crops in some regions of the world may benefit, the overall impact of climate change on global agriculture will be negative.

Localized extreme weather conditions will trigger pest and disease outbreaks, resulting crop loss and requires adaptable management responses (FAO-PAR 2011). Improving adaptation of the agricultural sector to adverse climatic conditions will be imperative to improving livelihood of the poor to ensure food security (FAO 2012). There are various options to make cropping resilient to climate change.

Sustainable Crop Production Intensification

The green revolution in India helped increased food production in India from 800 million tons to over 2.2 billion tons between 1961 and 2000, this increase due to the highly chemical centric farming took a heavy toll on the natural resources, notably, soil, as recently explained by Nair in his invitational chapter "India's tryst with the green revolution—A critical Analysis", in the book "Agriculture Research Updates" published by Nova Science Publishers. Ground water got depleted, excessively contaminated making it non potable, upsurge of pests and diseases, loss of biodiversity due to the rice–wheat monocropping, huge soil degradation, above all, even escalation of even cancer in human beings in Gurdaspur district in Punjab State "the cradle of India's green revolution", due to unbridled use of herbicides and pesticides, and the district received the dubious distinction of being called the "cancer capital of India".

A productive agriculture system that enhances and conserves natural resources through a dependable and sustainable ecosystem approach capitalizes on natural biological inputs and processes. It reduces the negative impacts on the environment and enhances natural capital and flow of ecosystem services. The climate smart/resilient agriculture practices focus on improving efficiencies and managing effectively biological processes.

The management practices include the following:

1. Maintaining a healthy soil
2. Cultivating a wider range of species and varieties in sole cropping, crop rotations and sequences
3. Using quality seeds and planting materials of well adapted, high yielding crop varieties
4. Adopting the integrated management of pests and diseases, also weeds
5. Managing water use efficiently
6. Biodiversity conservation
7. Agroforestry
8. Integrated farming system.

Both crop production and climate change adaptation and mitigation require resilient practices leading to the establishment of a resilient ecosystem. In terms of technologies, productive and sustainable agricultural ecosystems make the best and effective use of crop varieties, and livestock breeds, and their efficient agro-ecological and agronomic management (Beddington et al. 2012).

Cropping System Management

Monoculture with cash crops has become vulnerable to climate change, as clearly shown by the monocropping of rice–wheat in India, during the green revolution. Mixed cropping, diversified cropping and intercropping have become resilient to climate change.

Intercropping: Intercropping is the practice of growing two or more crops I the same patch of land, very popular in developing countries like India. The crops cultivated should not be competitive for space, sunlight, nutrients and water's. Normally, a shallow-rooted crop is intercropped with a deep-rooted one. Intercropping produces higher yield and land equivalent ratio (LER) than a sole or single crop. Yields in intercrops are more stable than that in sole crops. Some of the more significant types are mixed intercropping, row cropping, and relay cropping. In mixed intercropping, the component crops are mixed in a definite proportion and sown. Row cropping involves the component crops arranged in alternate rows. Variations include alley cropping, where they are sown between row of trees, and strip cropping, where multiple rows, or a strip, of one crop is alternated with multiple rows of another crop. Intercropping under scattered trees is the simplest form of agroforestry. The traditional intercropping systems consist of growing agricultural crops under scattered or systematically planted trees on farmlands. There are more diverse multi-storied home gardens in humid areas and less diverse, two-tired canopy configurations (trees crops) in drier areas.

Crop Rotation

Crop rotation is a practice of growing different types of crops on the same patch of land in a specific sequence in different seasons. Using some forms of crop rotation, farmer can reduce the need for the synthetic fertilizers as well as keep their fields under continuous production instead of letting them lie fallow. Crop rotation offers various soil fertility benefits to the soil, such as, soil enrichment of the nitrogen status of soil through symbiotic nitrogen fixation, especially when legumes plants are planted in the crop rotations, such as bean species, for example *Vigna unguiculata*, and red gam *Cajanus cajan*. Crop rotation invariably reduces crop pest attacks, through insects and pathogens, also reduce the adverse impact of climate change. Also, by requiring planting and harvesting at different times, allow more land to be farmed with the same amount of machinery and labor.

Crop Rotation and Adjustment of the Crop Variety

A sound crop production system needs to adapt to climate change and become climate resilient. Famers use resistant or early maturing crop varieties, practice mixed copping and plant trees to provide fodder and shade/shelter for their crops. The adverse impacts of climate change on grain legume crops can be mediated through suitable strategies, such as, using stress tolerant crop varieties and adjusting sowing time to favorable weather and mixed/inter cropping. Indigenous crop varieties which are drought resistant/tolerant can be grown.

Crop varieties resistant/tolerant to drought, flood, salinity, crop pests and other stress conditions, such as, biotic and abiotic conditions, ae constantly being evolved. Production of such crop seeds is necessary both in the informal and formal sectors. There are some such notable example in India, through farmers' initiative. At the same time, conservation of plant genetic resources can be made through in situ and ex situ methods and village gene and seed banks. Plant breeders must develop crop varieties to adapt to climate change. Participatory plant breeding is another approach where farmers and scientists (plant breeder) join hands in developing seeds to adapt to climate change.

Inclusion of Legumes into Cropping System

Legumes improve soil N status through biological N fixation and scavenging residual soil N, making it available for subsequent crops. Legumes are prolific leaf shedders, which act as a natural manure rich in nitrogen and help conserve soil moisture and maintain soil fertility. Inclusion of legumes into the cropping system as a sole/single crop, or as a mixed crop, relay or catch crop, a cover crop and a green manure crop, in mixed cropping systems and as a forage crop to improve livestock nutrition, and, all of them the can be climate resilient. Table 8.2 gives data on biological nitrogen fixation (BNF) by different legume plants.

Table 8.2 Estimated amount of N fixed by different legume plants as per data provided by FAO, Rome

Plant	Botanical name	N fixed (kg/ha/year)
Horse gram	*Vicia faba*	45–552
Pigeon pea	*Cajanus cajan*	168–280
Cowpea	*Vigna unguiculata*	73–354
Mung bean	*Vigna mungo*	63–342
Chickpea	*Cicer arietinum*	103
Lentil	*Lens esculenta*	88–114
Peanut	*Arachis hypogaea*	72–124
Beans	*Phaseolus vulgaris*	40–70
Pea	*Pisum sativum*	55–77

System of Rice Intensification (SRI)

SRI is the method of alternate wetting and drying of soil during the rice cropping season, rather than continuous flooding. Flooded rice consumes a lot more irrigation water and produce methane which is twenty five times more potent than carbon di oxide as a GHG. SRI is a resource conserving system of rice cultivation which gives better yield with less seed, less water and less nutrient need. However, it requires higher manpower need for cultivation, which, of course, is compensated by low input cost. The benefits of Sri are the following:

1. SRI increases crop production with lower inputs
2. It is produces productive and robust rice phenotypes from given plant genotypes
3. Under SRI system of rice cultivation, the rice crop is tolerant to both biotic and abiotic stresses and results in lower GHG emission
4. The system helps farmers to adapt to, and, mitigate adverse climate impacts.

Conservation Agriculture

Conservation agriculture is an approach which combines limiting soil disturbance to the minimum, maintaining soil cover, and diversifying crop production. It increases the resilience of the farming systems in a number of ways. The soil is kept fertile and protected from erosion and evaporation, water and plant nutrients are used efficiently, ground cover slows down the low of water on soil surface, allowing it to soak and infiltrate into the soil for subsequent crop use. Conservation agriculture reduces carbon di oxide emissions from tractor use, as the farm power requirements are lower and fewer passes only are needed across the fields. These “carbon savings” exceed the carbon costs related to the use of chemical herbicides for weed control. Soil cover reduces moisture loss by evaporation, maintains soil temperature, reduces soil erosion, restores soil carbon by decomposition of crop resides and provides food for soil microorganisms. Crop rotation reduces pest and disease attack and replenishes soil nutrients. Minimum tillage increases population of beneficial soil microorganisms in soil which takes care of soil tillage, to build up soil properties.

Conservation Agriculture constitutes of the following three components:

1. Reduced tillage
2. Retention of crop residues on soil surface
3. Crop diversification such as, rotation, intercropping (Table 8.3).

Integrated Nutrient Management (INM)

INM balances the use of organic matter obtained from the farm or from other sources, with the judicious use of nutrients from mineral sources. It is a holistic approach which aims at higher yields and improvement in soil fertility.

Table 8.3 A comparison between conventional agriculture and conservation agriculture

Properties	Conventional agriculture	Conservation agriculture
Soil structure	Repeated tillage reduces soil organic matter, destroys soil structure, develops hard pan	Minimum tillage does not disturb soil structure and helps build soil organic matter
Soil moisture	Evaporation more waterlogging due to hard pan	Reduced evaporation leads to higher infiltration
Soil erosion	Crust formation and sealing of surface pores prevents water seepage encouraging erosion. More erosion in sloppy land	Cover crop and mulch protects from rainfall impact reducing erosion, roots bind soil particles resisting soil erosion
Soil fertility	Plowing beaks up	Crop residues and cover crops add soil organic matter, compost addition enriches
	Organic matter, kills soil microorganisms affecting soil health. Monoculture removes valuable soil nutrients	Soil fertility
Pests and weeds	Monoculture encourages both pests and diseases requiring both pesticides and herbicides	Weeds smothered by cover crops, crop rotation reduce pest attack
Cost and labor	More labor and fuel required for conventional agriculture, but, chemical use harmful to environment	More labor in the first year, but, reduced cost in the following years

Apart from applying organic manures, and, oil cakes, nutrient recycling by legumes could be a potential component of INM. Green manuring with legumes, such as, *Gliricidia maculata* or *Crotalaria juncea* (Sunhemp) is a common practice in the rice growing regions of India. A 6–8 weeks old green manure crop of sesbania or Sunhemp accumulates about 3–4 tonnes of dry matter/ha and fixes 100–120 kg N/ha. This can supplement upto 50% of the N requirement of the rice crop.

There are two ways to manage nutrients: micro dosing and fertilizer application strategies. Micro dosing involves application of affordable quantities of fertilizer to the planting holes at the time of sowing/planting and top dressing 3–4 weeks later. However, micro dosing is a time consuming field operation. Good management of N fertilizer application includes application of fertilizer on the basis of soil tests. The blockbuster book by this author titled "Intelligent Soil Management for Sustainable Agriculture: The Nutrient Buffer Power Concept", discusses the salient aspects, especially with regard to a revolutionary fertilizer management technique, now globally known as "The Nutrient Buffer Power Concept".

Carbon Sequestration: Leaving the crop residues in the soil, without burning, minimum tillage and addition of compost, can sequester carbon in the soil. To encourage carbon sequestration, cover and catch cropping, restoration of degraded land, soil conservation and erosion control methods ae essential. The establishment of deep-rooted and perennial plants will facilitate carbon sequestration, and trees ca be used for live-fencing, windbreaks, fodder as riparian forest buffers etc.

Integrated Pest Management

Climate change often increases pest attack in crops. Indiscriminate use of pesticides though provide pest control for a short term, cause environmental hazards later on. Integrated Pest Management (IPM) is the solution. This is a broad ecological approach involving use of appropriate measures to discourage the development of pest population in the crop fields and keep them within economic injury level. It is a system that relies on varietal, cultural, mechanical, physical, biological and need-based chemical control tactics to maintain a natural balance in the crop system. The IPM principles include growing healthy crops, understanding the ecological processes in the fields, encouraging natural pest management mechanisms to maintain ecological balance among pest populations and their natural predators, such as parasites, antagonists etc.

Integrated Farming System (IFS)

IFS is a diverse system which includes crops, livestock or forestry which is integrated at various scales to enhance the efficiency and environmental sustainability of the different production methods. IFS utilizes different production enterprises, where the byproduct of one component is used as an input for another component. IFS provides additional adaptation options to climate change and produces diversified products, such as, grains, fruits, vegetables, egg, milk and meat.

Climate Resilient Approaches

At the field level, there wide range of practices and approaches which ca contribute to increased production, while still focusing on environmental sustainability. These practices and approaches for crop production can provide adaptation measures and/ or mitigation benefits. With appropriate technical, institutional, socio-economic and policy infrastructure in place, there is huge potential for crop management practices and approaches for mitigation of climate change.

Preservation of Genetic Resources

Crops, animal and other genetic resources will need to adapt to the impacts of climate change. Breeding activities should ensure that genetic resources for food and agriculture are able to adapt to new production conditions. The following are the characters important for climate change adaptation:

1. Capacity to tolerate high ambient temperatures and droughts
2. Resistance and tolerance to fire, especially for trees
3. Resistance to both insect pest and diseases attack
4. Capacity to utilize scarce and poor quality feed and soil
5. Tolerance to lower water quality, especially saline water, for aquatic organisms
6. Capacity to range over harsh terrain in search of feed and water in case of livestock.

Energy Management

Fossil fuels are likely to meet the bulk of primary requirements, price of which is escalating by the day, while the costs of renewable energy sources have been declining. With this developing trend, the gap between energy needs and energy access will be large. Agriculture and energy have always been closely inked over time. The energy generated by the agri-food system can be partially used in the food supply chain or exported outside the system.

Energy utilization has increased over the years in farm mechanization, fertilizer and food processing and transportation. Since prices of nitrogenous fertilizers and other fossil-fuel dependent ones have escalated over the years, the agri-food systems dependent on fossil fuels pose serious challenges to agricultural development, and food security. Food losses occur at all the stages of the supply chain. It is a challenge to develop a food production system which emits les of green house gases and at the same time be resilient to fluctuating energy costs. This calls for an energy efficient/ smart system which must ensure the following:

1. Improve energy use efficiency at all stages of the food production chain, starting from the field to the factory
2. Use diverse energy sources with an emphasis on renewable energy and contribute to renewable energy production through integrated food and renewable energy production
3. Require improved access to modern energy services.

Bioenergy is used primarily for cooking and heating which may jeopardize food security and harm the environment, if not well managed.

The energy sector, which produces nearly 60% of carbon di oxide emissions, is the largest contributor to climate change (FAO 2011). Globally, primary farm and fishery production accounts for around 20% of the total energy demand for food, but, produces about 67% of green house gases (FAO 2011). Against the background of high and volatile fuel prices, energy-smart food systems, which improve access to modern energy services and increase energy diversity, can contribute to energy security. The energy supply needs to be "climate-proofed" as much as possible to ensure that energy use in the agri-food system can be climate-resilient. Precision agriculture, including a more efficient fertilizer use, will lower carbon di oxide and nitrous oxide emissions.

Inefficient use of nitrogenous fertilizers leads to losses that are usually above 50% and sometimes can even be as much as 60–70% (Cassman et al. 2002). Increasing the efficiency of fertilizer application can result in optimal plant growth with minimal inputs, improving energy balance in food production.

Soil tillage, concerning land preparation like plowing, is the single most energy consuming operation in a cropping cycle. Minimum tillage or no till system has gained importance in this connection. Practices such as zero tillage used in conservation agriculture have the potential to bring to bring about saving in energy conservation to the extent of 40–50% (SCCA 2012). For grain drying, modern continuous flow dryers can be used with much lower levels of energy than conventional dryers. Reduced energy use can also be achieved through insulation of dryers, the re-circulation of heat recovery of out-going air and improved instrumentation and automatic control. Solar power (photovoltaic or solar heaters), wind and geothermal energy are all sources of energy, in particular, suitable for remote rural areas (see Figure).

Climate Resilient Livestock and Fishery

Increased ambient temperature, and, precipitation, both, affect livestock. Indirect impacts are modifications in ecosystems, changes in the yields, quality and type of feed crops, possible increases in animal diseases and increased competition for resources. At the same time, livestock food chain are major contributors to greenhouse gases (GHGs) (FAO 2006).

The principal strategies for climate resilient livestock system are the following:

1. Improved grazing system through rotational grazing to reduce methane emissions and carbon di oxide removals through soil carbon sequestration
2. Efficient manure management
3. Efficient pasture management and nutrition
4. Animal breeding to select more productive animals with lower methane emission capacity and higher resistance to climate change
5. Better nutrition and health management
6. Early warning systems
7. Agro forestry practices
8. Mixed farming systems.

Ecosystem approach to fisheries and aquaculture would integrate fisheries, aquaculture, aquaculture and agroforestry. In rice-fish culture in flooded rice fields there is synergetic increase in production of both rice and fish. Fish harvest is done by reducing water level, forcing them to swim to the lowest-lying part of the fields, where they can be caught easily. The agricultural producers get two harvests—rice and fish—from the same land, enhancing the farmers' income and ensuring food security.

Climate Resilient Forestry

Forests and trees provide a range of forest products and ecosystem services. Mainstreaming climate change into forest policy and practice will entail capturing synergies and managing trade-offs with other forest management objectives. Climate resilient agriculture in the forest sector will entail designing adaptation actions targeted at the most vulnerable communities and sectors of the population and forest systems. Climate change responses, in particular, mitigation policies, are leading to changes in markets, for the forest products and may drive substitution of forest products for the production of more energy or carbon-intensive products. Forest planning and management will have to be sensitive and adjust to such market changes.

The International Council for Research in Agroforestry in Nairobi, Kenya, (ICRAF), currently renamed as World Agroforest Center, defines agroforestry "as a farming system that integrates crops and livestock with trees and shrubs". Agroforestry practices such as use of shade trees are another example of an integrated practice. The resulting biological interactions in the agro-forestry system provide multiple benefits, including diversified income sources, increased biological production, better water quality, and improved habitat for both humans and wildlife. Agroforestry has multiple benefits such as increase in carbon sequestration, higher income from non-wood and timber materials, controlling soil erosion and maintaining soil fertility. Agroforestry brings in a number of benefits in terms of soil carbon sequestration.

Biodiversity Conservation

Biodiversity is especially important in a climate-resilient approach, because, it contributes to pest and disease management, which has direct effect on yields and revenues and can be very costly and labor intensive. Agriculture and allied agricultural production systems depend on the following three components:

1. The diversity of the species
2. The diversity within each species
3. The diversity of ecosystems.

Greater diversity of crops and other living organisms in a cropping system is required to ensure farm resilience to climate changes, ensure economic profitability and stability.

Integrating multipurpose crop varieties, which can improve the functional and productive management on the farm and are climate resilient.

References

Beddington, J., Asaduzzaman, M., Clark, M., Fernandez, A., Guillou, M., Jahn, M., Erda, I., Mamo, T., Van Bo, N., Nobre, C.A., Scholes, R., Sharma, R., and Wakhungu, J. 2012. Achieving food security in the face of climate change. Final report from the Commission on Sustainable Agriculture and Climate Change, Copenhagen

Brown, S., and Lugo, E. 1990. Tropical secondary forests. Journal of Tropical Ecology, 6, 1–32

Cassman, K.G., Dobermann, A., and Waters, D.T. 2002. Agroecosystems, nitrogen use efficiency, and nitrogen management. Ambio, 31, 131–140

FAO 2006. Livestock's long shadow: environmental issues and options. Rome (available at ftp://ftp.fao.org/docrep/FAO/010/AO701EOO.pdf)

FAO 2011. "Energy - Smart" food for people and climate. Issue paper (available at http://www.fao.org/docrep/014/i2454e00.pdf)

FAO 2012. Economics of plant genetic resource management for adaptation to climate change. A review of selected literature by S. Asfaw and L. Lipper (available at http://www.fao.org/docrep/015/an649e/an649/an649e00.pdf)

FAO-PAR (Platform for Agro-biodiversity). 2011. Biodiversity for food and agriculture (available at http://agrobioiversityplatform.org/files/2011/04/PAR-FAO-book_Ir.pdf)

Farage, P., Ardo, J., Olsson, L., Rienzi, E., Ball, A., and Pretty, J. 2007. The potential for soil carbon sequestration in three tropical dryland farming systems of Africa and Latin America: a modeling approach. Soil and Tillage Research, 94(2), 457–472

Macdonald, A.J., Powlson, D.S., Poulton, P.R., and Jenkinson, D.S. 1989. Unused fertilizer nitrogen in arable soils – its contribution to nitrate leaching. Journal of the Science of Food and Agriculture, 46, 407–419

Noellemeyer, E., Frank, F., Alvarez, C., Morazzo, G., and Quiroga, A. 2008. Carbon contents and aggregation related to soil physical and biological properties under land-use sequence in the semiarid region of Central Argentina. Soil & Tillage Research, 99, 179–190

Soil Conservation Council of Canada (SCCA). 2012. Producers are finding the reduced tillage zone (available at http://www.soilcc.ca/ggmp_region/nb-02php)

Sparkes, J., and Stoutjesdijk, P. 2011. Biochar: implications for agricultural productivity. ABARES technical report. 11(6). Australian Bureau

Turral, H., Burke, J., and Faures, J.M. 2011. Climate change, water and food security. FAO Water Report. No 36, Rome, FAO (available at http://www.unwater.org/downloads/unw_ccpolweb.pdf)

Chapter 9
Climate Resilient Village

Application of Climate-Resilient Practices

Agricultural production systems are expected to change in response to climate change to ensure food security. Climate variation may adversely affect food production in the future. To meet such challenges, there are many options to reduce the negative impacts of climate change on agricultural systems, make them resilient to climate change, and reduce emissions of GHGs (green house gases). Many technologies and practices increase crop yields, farm income, and, input-use efficiency, and will also reduce emissions of GHGs. Climate resilient agriculture (CRA) is suggested to address the adverse effects of climate change and food security. It is intended to achieve the three objectives of ensuring crop productivity, adapting and building resilience and reducing emissions of GHGs (Green House Gases).

What Is the Concept of CRA Approach?

The CGIAR (Consultative Group for International Agricultural Research) research program for climate change, agriculture and food security (CCAFS) has developed the idea of climate-resilient villages (CRA) approach as a means for effective agricultural research and development (AR4D) with the following objectives:

1. Sites for testing, through participatory methods, technological and institutional options.
2. Sites where climate change in it's broadest context is considered, but in relation to local realities; long-term adaptation, avoiding maladaptation, climate risk management and low emission development.
3. Embodies a holistic vision for climate change action—not a silver bullet approach.
4. A platform for socially inclusive, multi-stakeholder collaborative work.

K. P. Nair, *Climate Resilient Sustainable Agriculture*,
https://doi.org/10.1007/978-3-031-84166-8_9

5. Founded on the principle of bringing CRA to scale.
6. Links local and global knowledge.

Practices, technologies, services and institutional options in a broad sense constitute the CRA approach. The Climate Resilient Village AR4D approach considers the following components in the proposed sites (see Figure). In AR4D approach, Climate resilient Village (CRVs) are multi-stake holder learning programs, participatory test-beds to generate evidences for further scaling.

The decision support to farmers and other stakeholders provided by an ideal CRV approach are the following:

1. Village/community agricultural land use plans and contingency plans considering current and future climate risks, agro-ecological and socioeconomic conditions, and, markets.
2. Portfolios of agricultural practices, technologies, and programs which do not become maladaptive in future climate and market scenarios are assessed using models.
3. Strategic guidance before and during the planting season on the most suitable CRA practices, technologies, services, processes, and institutional options considering markets and resource availability, such as, capital, labor, and marketing done in a participatory mode with local farmer group, such as, self-help groups (SHGs), water users association (WUA), local extension systems, or local private-sector actors.
4. The tactical guidance to farmers constitute using real-time weather forecasts and value-added information and communication technology (ICT)-based agro-advisories. Further, on accessing good quality inputs and technologies for improving water-nutrient-energy use efficiencies and on risk transfer through insurance mechanisms in case of crop failure and livestock losses. On the nutrient use "**The Nutrient Buffer Power Concept**", developed by this author will come in handy.
5. Policy level guidance on barriers and options to unlock CRA with local and national development, including, considerations of financial needs to drive scaling. The CRA approach differs from the existing agricultural development approaches, which, normally, depend on substantial external funding.

Adaptation promotion and transformation options locally with capacity building entail the CRV approach. The approach aims to have Aiming positive impact on agriculture-dependent communities is the strategy of this approach. Climate Resilient Village V AR4D sites could be a cluster of villages or landscapes with one or many villages.

What Are the Broad Objectives of Climate Resilient Agriculture?

In principle, there are three objectives of the climate resilient agriculture, which are the following:

1. Food security: Since climate change affects four dimensions of food security, which are, availability, accessibility, absorption and stability, the major objectives of CRV is to ensure food security by adopting sustainable agricultural practices.
2. Adaptation: The CRV seeks to increase the capability of the stakeholders by building social capital, economic resources, awareness and training and adopting climate resilient technologies and developing infrastructure against climate change.
3. Mitigation: Mitigation strategies to reduce the emission of green house gases (GHGs) from agricultural operations and allied sectors are attended to by CRV.

The climate resilient village (CRV) interventions include the following five points:

1. Weather-smart,
2. Water-smart,
3. Seed/breed (both crop and livestock)-smart,
4. Carbon/nutrient-smart, and,
5. Knowledge-smart (Table 9.1).

Table 9.1 Which are the climate resilient village interventions

Details	Interventions
Weather-smart	Weather forecasts, climate informed agro-advisories, weather insurance, climate analogues, as a tool for forward planning, strategies to avoid maladaptation
Water-smart	Aquifer-recharge, rainwater harvesting, community management of water, laser-land leveling, micro-irrigation, raised bed planting, solar pumps
Seed and breed-smart	Adapted varieties of seeds, breeds (livestock), seed banks, including community-based activities
Carbon/ nutrient-smart	Agroforestry, minimum tillage, land use systems, use of leaf color charts, GreenSeeker, livestock management, integrated nutrient management, biofuels
Energy-smart	Bio fuel, fuel efficient engine, residue management, minimum tillage, solar solutions
Knowledge-smart	Cross-sectoral links, local institutions, including learning platforms or farmer-to-farmer learning and capacity development, contingency planning, financial services, market information, gender equitable approaches, and off-farm risk management strategies

Implementation of CRV

What Are the Principles for Effective Implementation of CRV Climate Resilient Agriculture (CRV)?

The effective implementation of climate resilient agriculture requires an integrated approach in which, science, technology, and decision making processes intelligently and pro actively interact with local socioeconomic conditions and cultures. A CRV is a platform for action research to test, to develop and to support the scaling of CRV technologies and approaches to help in deriving practical adaptation options to ensure food security, improve resilience, and decrease emissions of green house gases (GHG). The CRVs are meant to generate and scale CRA technologies and social learning approaches or to scale CRV methodologies. It aims at identifying the best climate smart agricultural options to draw out lessons for policy makers and agricultural practitioners. The testing is done through a multi-stakeholder collaborative platform at CRV sites like a cluster of villages, small landscapes, or 10 km^2 grids as every CRV site has its own theory of change, liked to national priorities to ensure that it is consistent with initiatives and actions across different scales.

Which Are the Four Most Important Strategies to Achieve Climate Resilient Villages (CRVs)?

1. An understanding of the effectiveness of a variety of CRA options (practices, technologies, services, programs, and policies) not only to enhance crop productivity and raise farmers' income, but also, to achieve climate resilience, enhance adaptive capacity, and wherever possible, reduce GHG emissions.
2. Developing solutions in anticipation of future climate change impacts.
3. An understanding of the socio-economic, gender, and biophysical constraints and enablers for adoption, and finally.
4. Testing and identifying successful adoption incentives, finance opportunities, institutional arrangements, and scaling out/up mechanisms while ensuring alignment with local and national knowledge, institutions, and development plans.

Stopped Here

The strategy of CRV approach is not only to enhance productivity and raise incomes, but also, to build climate resilience and reduce emission of green house gases (GHGs).

There are many options and they differ from one another based on the CRV site, its agro-ecological characteristics, the level of development, and capacity and interest of the farmers and of the local government. An ideal CRV approach provides five types of decision support to farmers and other stakeholders. They are the following:

1. Village/community agricultural land use plans and contingency plans considering current and future climate risks, agro-ecological and socio-economic conditions and markets.

2. Portfolios of agricultural practices, technologies, and programs which do not become maladaptive in future climate and market scenarios which are assessed by using models.
3. Strategies guidance before (where feasible based on seasonal forecasts) and during the planting season on the most suitable CRA practices, technologies, services, processes, and institutional options, considering market and resource availability, such as, capital, labor, and markets. This is done in a participatory mode with local farmer groups, such as, farmers' self help groups (SHGs), water users' associations (WUAs), local extension systems, r, local private-sector actors.
4. Tactical guidance to farmers on using real-time weather forecasts and value-added information and communication technology (ICT)-based agro-advisories; on accessing good quality inputs and technologies for improving water-nutrient/energy use efficiencies.
5. Policy level guidance on barriers and options to unlock CRA with local and national development, including consideration of financial needs to drive up scaling.

The key focus in a typical CRV is on strengthening capabilities and empowering the farming communities and their local organizations. In the process, partner in the CRV approach commit only limited resource support to catalyze the process.

What Are the Steps Needed to Implement the CRV Strategy?

The CRV approach consists of the four following steps:

The CRV approach guides adaptation planning and on the ground implementation, with the objective of transforming farming communities to be more climate smart. The four steps involved are the following:

1. Baseline Assessment, Including Climate-Risk Analysis and Gender and Social Inclusion Analysis

Historical climate data are analyzed to assess risks, such as, the probability of sowing failure due to inadequate rainfall at the start of the season. It also includes risks of wet and dry spells, and periods of heat or cold stresses during the cropping season. Future climate scenarios are analyzed to assess the likely change in risk profiles and long-term suitability of the main cropping and livestock systems. Secondary data are used to assess GHG emissions from agriculture and simple calculators in those regions, where mitigatory steps are warranted/relevant.

In baseline assessment, the natural resources (land, water and soil) and availability of physical resources (infrastructure, such as irrigation, technology and markets) are assessed. It also includes assessment of socio-economic resources such as human (labor and education), financial (source and amount of revenue and budget) and social resources (institutions and networks). Local communities help guide CRV design process, which includes their knowledge of climate risk management.

A quick participatory assessment is made of the different practices and technologies being used in the village/region, as well as the potential ones which can be adopted, guided by agronomic investigations in the region or homologous regions elsewhere. Availability of climate informed services, such as, weather-based agro-advisories and agriculture insurance schemes in the region ae also assessed.

Assessment of current adaptation policies and institutions at different scales is essential to identify the potential for Cooperation with the leveraging support from other entities. These assessments indicate the scope for mainstreaming the CRV approach into local adaptation policies and plans, including other development initiatives. An evaluation is also done by gender and age to understand the different perspectives, division of labor, decision making patterns, opportunities, and capacities in households and communities.

2. CRV Design

Identification and prioritization of climate-smart technologies, practices and services based on biophysical, socio-economic, gender, policy and institutional context and possible synergies and trade-offs amongst individual activities, are considered for CRV design. This step primarily focuses on developing a portfolio of practices and technologies dealing with food security, adaptation, and mitigation and on-climate information services which need to be initially piloted in the CRVs. The portfolio includes the six climate-smart technologies already explained earlier in this chapter.

Strategic decision making in the CRV approach is a participatory activity. Proposed interventions must be relevant to the conditions defined by the baseline, explained earlier in this chapter, and show promise for wider relevancy of scale beyond the CRV. The CRA practices, technologies and services included in the portfolio for any specific site are the outcome of the group discussions with both men and women farmers, local government officials and researchers. Various tools, such as, choice of experiments, indicator-based ranking methods, and simple stakeholders' prioritization, are used for this exercise. The CRA portfolio may be influenced by gender difference in awareness levels, as well as, b the choice of CRA technologies by the stake holders. Several CCAFS projects have published details of how CRV portfolios of climate-smart interventions have been selected and designed.

3. Creating Evidence

Evaluation and development of portfolios of climate-smart interventions are done to create evidences. This is a two-step process. First, a narrow-range of interventions/portfolios is evaluated in a multi-stake holder research platform, such as the CGIAR (Consultative Group for International Agricultural Research—explained earlier in this book), national agricultural research systems, NGOs, private-sector players, farmers and their institutions, established in CRV sites. Assessment of benefits, synergies, and technologies is made from the individual farmers and the concerned community. The detailed evaluation is made by surveys, farmers' groups, and ICT (Information and Communication Technology) based feedback system.

Modeling is often used to supplement the results of on-farm evaluation, especially to estimate the benefits of resilience and mitigation with passage of time. In the second step, the final portfolios are evaluated by the concerned farmers on a larger scale. Crowd-sourcing of experiences is used in order to understand farmers' preferences and technology adaptation domains across CRV sites in the concerned region. Due consideration is also given to the identification and evaluation of constraints and barriers to the adoption of interventions, along with ways to overcome the hurdles.

4. Scaling

This involves scaling up through policies and institutions, and scaling out to large areas through farm-to-farm ad ICT-based approaches. After generations of evidence the promising innovations are scaled up. Scaling mechanisms tested across the regions include the following:

1. Horizontal scaling (scaling out) of climate-smart options. CRVs provide demonstration sites for farmer-to-farmer learning (through self help groups or producer organizations) and or/enable local promotion of CRA options through local government plans, programs, and policies or through private sector business models.
2. Vertical scaling (scaling up): CRV research and lessons learned provide evidence for the efficacy of practices, technologies, services, processes, and institutional options and are able to influence effects/results on a large scale.

Horizontal scaling out is done from farmer to farmer through networking using ICT-based tools, farmers' fairs, farmer exchange visits and participatory videos of the adopted technologies, or through engagement with local government institutions. Connecting with women's organizations is another effective scaling-out approach. Institutional and financial needs are determined, for both local and national (or subnational) contexts, often using appropriate business models. The potential for consolidation and convergence of current agricultural development schemes/plans in the region is explored. Consideration of national and regional gender machineries and social development programs also helps integrate social inclusion concerns into CRVs as well as to orient social development program to climate change impacts and action.

Monitoring and evaluation provide feedback to the CRV design process, which is based on learning and adjustment. Finally, adaptation pathways for CRA approach—including technology and services portfolios, capacity building programs, institutional and financial mechanisms, and, potential costs of program implementation—are presented to policy makers for scaling through replication in other sites, or, provision of incentives for adoption through large-scale agricultural investment plans. Although conceptually, this step is placed at the end of the sequence, it can vary according to the CRV context.

The following steps are involved in setting up the CRVs:

Step 1: Define the purpose and scope of the CRV which is being established

To begin with, discussion has to take place on climate change impacts, climate resilient agriculture, concepts and practices including key components and elements of CRV approach and ways to achieve multiple outcomes—resilience building, increasing adaptive capacities and productivity, reducing gender inequality, and social inclusion. At the program/national level, a framework which justifies the need for a CRV approach needs to be ensured. During this step, discussion has to take place on CRA, CRV and climate change adaptation to address climate change and food and nutrition security, building climate resilience, increasing adaptive capacities and productivity of agriculture centers, livelihoods, gender and social inclusion.

Step 2: Identify the climate risks in the target region

In order to understand the climate risks, and vulnerabilities, employing available assessment tools such as participatory rural appraisal (PRA) will be used in the target area (agro-ecological regions, districts). The current and anticipated impacts of climate change on local agriculture (crops, livestock etc.), and the associated landscapes, can be understood by conducting rapid baseline study, household-level assessments, and focus on group discussions etc. Location specific climate risk maps may be constructed for vulnerability assessment. After deciding the PRA tools to be used, the PRA exercises have to be conducted by the community along with the baseline survey.

Step 3: Locate a CRV in a small landscape

A CRV should represent specific agro-climatic conditions and climate risks which can co-exist of a single village or cluster of villages, preferably associated with small landscapes and watersheds. Within the target region, it is required to clearly identify and delineate geographic areas of the proposed CRV location (village, community, and landscape). Ethnic or cultural diversity and different farmer types should be considered and recognized. CRV-specific climate risk mapping investigations and livelihood analysis can be undertaken using simple PRA tools and focus on group discussions. After undertaking all these processes, the CRV is to be decided.

Step 4: Organize a stakeholder workshop

A stakeholder workshop has to be organized to agree on the objectives, purpose, and scope, of the CRV; identify and agree on partners, resource institution and lead entity; plan early for scaling; prepare a three years work plan.

Step 5: Evaluate CRV options

Through participatory action research, evaluate CRA options addressing climate impacts and improving productivity and sustainability. In evaluating CRA options, recognize that CRV is a platform for testing, deriving, and subsequently promoting locally-adapted options in out scaling efforts. It is required to prioritize options unique to the local contexts of CRV.

Step 6: Develop a portfolio

CRA options in crop management practices and technologies for different target groups and agro-ecological situations need to be integrated. While developing a CRA portfolio, the climate risks and livelihoods and expectations of the target groups must be considered. During the actual field visit, ensure there is exchange of ideas between/among participants and CRV stakeholders. At the end of each visit, facilitate reflections sessions. The key lessons are to be synthesized and farm-level adaptation strategies are to be identified using field-derived lessons and emerging ideas.

Step 7: Identify scaling opportunities and pathways

In consultation with the stakeholders, a plan has to be prepared to scale-up, identify opportunities for horizontal and vertical scaling, organize events to share CRA technologies and tools and to disseminate the findings, and engage media and communication specialists in promoting successes and stories.

Step 8: Monitor and Evaluate (M&E) uptake and outcome

The required activities are to implement M&E of uptake and outcome: conducting (on annual basis) local reviews and assessment: assessing outcomes from action research undertaken in RV on a regular basis, using multidisciplinary teams, multiple stakeholder representatives, and community representatives; using indicators, such as, productivity, gender, livelihoods, food and nutrition security, and resilience to monitor progress; and developing outcome stories (substantiated by evidence and demonstrated by the scale of adoption, with the likelihood of being sustained).

Participatory Vulnerability Assessment (PVA)

PVA is an approach which combines various tools and methods which communities can use to generate and analyze information enabling them to identify options to address current climate change impacts and adaptation options. In practice, PVA outputs include community profile (village history, current state of natural resources, land use, tenurial arrangements), livelihood profile, climate change perceptions and coping mechanisms (perceived climate change impacts). Data gathering methods may include focus group discussion with PRA tolls, key informant and group interviews and review of secondary data and climate-related documents.

Participatory Action Research (PAR)

PAR, an approach which brings together, farmers, researchers and resource institutions, can produce sustainable solutions to address climate change impact in agriculture. PAR as defined by CCAFS is a "collaborative approach to research, education and action, and helps gather information in order to solve social or environmental issues". PAR in the context of resilient building in agriculture is used to test, identify and evaluate technological options and processes. Unlike conventional action research, testing and development of technologies are done in the farmers' fields. These technological options, lessons and ne knowledge can be scaled out through indigenous and emerging pathways.

Assessment of Statistical Analysis and Needs

In order to make a situation analysis, the following information is to be gathered by using participatory methods, such as, focus group discussion (FGD), Key Informant's Interview (KII), and collection of information from secondary sources. The following aspects need to be considered:

1. Geographical location
2. Demographics
3. Local climatic information
4. Natural resource utilization (land use, water resources, mining)
5. Production and livelihood systems
6. Food security status and trends
7. Hazards and vulnerability
8. Mitigation measures
9. Climate change perception
10. Climate change impacts
11. Institutional landscape and governance
12. Natural resource management initiatives
13. Organizational landscape
14. Information network
15. Social and gender differentiation
16. Health/nutrition profiles and other livelihood outcomes.

Participatory Rural Appraisal (PRA) Tools

Participatory Rural Appraisal that facilitates analysis of community situations and problems is undertaken by the community members themselves. Selected PRA tools to be used are outlined in Table 9.2.

Table 9.2 Selected PRA tools

Name of PRA tool to be used	Purpose
Community mapping	To identify community's boundaries, road network, river and springs, landmarks, infrastructure, houses
Historical time-line	To take stock of the significant events such as major disasters, socio-economic and political events, changes on land scape, and natural resources. It also shows the trends, frequency and intensity of events
Seasonality calendar	To depict pattern of climate, livelihood activities, occurrence of pests and diseases, health issues, social and cultural activities in different months of the year
Livelihood matrix	To list various works and identify each work performed by men or women. It also shows who mostly will be affected if a climate hazard adversely impacts livelihood
Problem tree	To identify three major livelihood problems and the reasons for the same; information and data gathered are rooted mainly on developmental issues; changes in patterns and those problems heightened by climate change
Vulnerability mapping	Identify and list vulnerable individuals and families and context analysis of socio-economic patterns useful to understand vulnerability in multiple risks including climate risks
Livelihood analysis	To know how individuals, households and groups of households make their living and the access of men and women to resources and services
Stakeholder analysis	To identify different stakeholders for every adaptation and mitigation activity proposed

Services for CRA

Under the CRA projects and programs, the services such as supply of inputs and technology transfer through extension services are provided to the farmers. The inputs such as seeds of different crops are provided by the input dealers at subsidized rates and seedlings of fruit trees and vegetables are provided by both private and governmental agencies to grow under general cultivation and poly-house technology. Similarly, other services such as soil testing facilities, issue of soil health cards, construction of micro-irrigation projects and rain water harvesting structures are made available to the farmers. Under the crop insurance scheme, loan farmers are supported to pay the premium for the insurance policy. Other services such as training at different levels, field demonstrations, exposure visits by farmers to new technology for resilient agriculture, exhibitions and *kisan melas* (farmers' fairs), field days, monitoring and evaluation of all these activities under these programs are provided to the farmers with the collaboration of the field extension staff. The summary of services provided for the farmers are given in Table 9.3.

Table 9.3 Details of services and service providers to the famers

Component	Services provided	Service providers	Beneficiaries
Input	Supply of seeds, agro chemicals, seedlings, plant grafts, traps, water harvesting structures, farm implements, soil testing, supply health cards, crop insurance etc.	Traders, input dealers State agriculture university, agriculture department, soil testing laboratories, soil conservation wings, banks etc.	Farmers
Extension service	Training, demonstration, exposure visits, organizing exhibitions, *kisan melas*, field days, farmer field schools, awareness camps, field monitoring and evaluation of programs		Farmers, self help groups

Barriers in the Adaptation of CRA Practices

There are several barriers in the adoption of CRA practices at the farmers' field level. The important barriers which need to be addressed are the following:

1. Lack of space on the small land holdings to install water harvesting structures, in particular, in rainfed areas.
2. Lack of labor availability to carry out the construction of the irrigation structures and taking out various agronomic practices.
3. Inaccessibility of good quality inputs by the famers from the input dealers and traders.
4. As agriculture is only the secondary occupation of the farmer, in some cases, the lack of interest acts as the barrier in the implementation of CRA practices.
5. Poor market facilities, especially for fruits and vegetables grown under CRA programs.
6. Lack of access to credit facilities for crop insurance of important crops.
7. Conflict of interest which can exist among farmers regarding the decisions to take the extension activities, such as, field demonstrations, training camps at village level, which may lead to both social and political confrontations.

The Role of Extension

The principal objectives of agricultural extension activities are to provide a system in which knowledge, skills and attitudes relating to farming practices are transmitted to the farming community through various channels and methods. The main aims are the following:

1. Transfer of technology and knowledge regarding improved agricultural practices and allied sectors.
2. To ensure practical application of knowledge and technology in the farm.
3. To improve living standards of farm families.

Choice of extension approach can influence the ability of extension services to contribute to food security and income, adaptation and resilience, and climate change mitigation. The most popular extension approaches used in CRA programs are the following:

1. Climate awareness programs/campaigns and exhibitions,
2. Climate trainings,
3. Climate workshops,
4. Plant health rallies,
5. Farmers field schools,
6. Exposure visits,
7. Demonstration of new farm technologies,
8. Participatory crop planning,
9. Dissemination of climate resilient technologies,
10. Establishment of "agro clinics".

Role of Rural Advisory Services

The expected effects of global climate change, which embrace higher temperatures, additional extreme weather events, water shortages, rising ocean levels, land degradation, disruption of ecosystems, consequently loss of biodiversity, will seriously compromise the ability of agriculture adequately the foremost vulnerable global population and impede the progress towards the reduction of hunger, nutrient deficiency disease among humankind, and poverty. It is vitally important to enhance the capability of farmer, globally, both rich and poor and marginal, to manage the risks of climate change and adopt effective measures to mitigate it. Promoting CRA agriculture practices involves strategies and agriculture practice which support climate resilient agriculture. Rural advisory service serves as a link between farmers and new sources of information and he approaches and methods to be sued in rural advisory service are demonstrations, trainings, individual farm visits, training of lead farmers and staff, training of input service providers, group discussions, exposure visits, farmer field schools, plant (agro) clinics, ICT (Information Communication Technology) based messaging and field days etc. Some of the important methods are described below:

1. Climate Awareness Mass Media Campaigns: Due to limitation of extension staff, extension can be more efficiently performed using mass media. The most preferred channels ae radio, television, telephone, SMS and APPs.

2. Climate Training: Education/training of intermediaries/extension personnel is important to update their knowledge related to climate change, their impacts, and consequences or on different adaptation and mitigation strategies. There are various topics in which training can be imparted. Farmer to farmer training can be organized through farmer field schools (FFS).
3. Plant (Agro) Clinics: This is an innovative approach which plays a vital role I timely diagnosis assuring food security. Timely diagnosis of both plant pathogen and insect attack on crops and advocating mitigation practices are vital to ensure crop security. In true sense, mobile plant clinics implies, a mobile clinic van with basic tools, public address system, relevant films with experts/plant "doctors", rushing to affected areas speedily and managing remedial measures to save the crop. Though the major role of plant clinic lies in diagnosis and advisory, the activities of plant clinics extend beyond crops, with emphasis on extension, working more closely with farmers and organizations involved in promoting food production.
4. Plant Health Rally Approach: Plant health rally aims to deliver customized crop health messages to farmers in locations where they assemble. The extension staff are trained on the best methods to quickly and effectively deliver the message and to answer farmers' questions. The recommendations to farmers are based on up-to-date information and evidence from research. The rallies also enable staff from crop disease and pest surveillance areas to work and share knowledge with extension staff for the benefit of the farmers.

It is an extension method to quickly rise awareness about major agricultural risks or threats on important crops, to promote the use of improved agricultural practices. They are usually held in public spaces by the local extension staff/officers and are open to everyone for attendance. A plant health rally may be spontaneous, attracting people with a banner and other announcements, or, may target farmers who have been specifically mobilized for the meet/event.

Contingency Crop Planning: When there weather aberration, contingency of growing another crop in place of the normally grown ones, arises. Depending upon the nature of weather change, the crop in question, is selected. Contingency cropping is highly location specific due to variation in the quantum and distribution pattern of rainfall. Crops have to be selected keeping in view its duration, to coincide with the length of the crop growing season. Generally, short duration crops or catch crops may suit the particular situation. Farmers with good economic means and motivation for achieving high profits with some amount of risk can go for crops of long duration. The long duration crops with flexibility or elasticity of yield are more preferable in contingent cropping.

It is participatory in nature, because, the local knowledge of farmers about the crops or various agricultural practices involved, and, the scientific knowledge of scientists, both play an important role in crop planning. It is prepared with the help of agro-meteorological department. Use of mobile phones, videos, radios etc., is done to address the issue of climate change by creating awareness among the farmers about the availability of different adaptation and mitigation strategies.

Climate Farmers Field Schools (CFFS): The Farer Field School (FFS) approach in an alternative extension method to top-down extension activities. The Farmer Field School (FFS) is a participatory non-formal extension approach based on experimental learning that puts farmers and their demands at the center (FAO 2002).

A group of farmers (25–30) assemble in the field school once a week. In groups of five, they observe and compare two plots over the course of an entire cropping season, right from the time of sowing up to harvesting. One plot is a control plot, where the crop is grown according to the conventional method, while the other demonstrates the possible "best practices". They critically observe the plant development, taking samples of insects, weeds and diseased plants, and constructing simple cage experiments or compare characteristics of different soils. The groups make presentation, basing on their observations at the end of the week, in a plenary session. In this program, farmers are able to investigate a wide range of topics relating to field problems and solutions. At the end of the season, a typical FFS group holds a "Field Day" to show the results to all the farmers involved in the program.

Farmer-to-Farmer Extension (F2FE): Farmer-to-Farmer Extension (F2F) approach in agriculture is a systematic communication between farmers. This approach aims to strengthen the information flow through farmers. As a generic term, we use the "farmer to farmer" term to communicate the message to other farmers. F2F approach helps to build an effective, farmer-centered extension system. It can reduce the implementation cost of extension services, also, the workload of extension staff. This approach empowers farmers as "change agents" and help extend adoption of different improved agricultural practices because farmers ae more willing to find out information from their colleagues d rather than the extension staff (Franzel et al. 2015). The farmers and local institutions, for example, producer organizations or village leaders, should play a key role in selecting farmer trainers and monitor and evaluate their input. This helps make the programs more accountable to the community of farmers or the groups they serve.

ICT (Information Communication Technology) Supported Network: ICT plays a crucial role as a medium of data and communication in global climate change awareness, adaptation and mitigation strategies. However, the provision and adoption of ICTs vary between areas where it is operated, developed and developing countries, urban and rural areas, and within rural areas themselves. The use of mobile phones, videos, radios etc., is done to deal with the difficulty of global climate change by creating awareness among the farmers about the supply of various adaptation and mitigation strategies.

Kerala's Killer Landslides—The Untold Story

On July 30, Wayanad district of Kerala received 6% of its annual rainfall in just a few hours leading to heavy landslides where more than 500 lives were lost and also resulted in huge property loss. It was a global tragedy of no comparable magnitude anywhere else. The landslides was the result of years of relentless exploitation of an ecologically fragile region—a hand-in-glove operation by the state government and the corrupt quarry owners. The disaster, can be traced back to activities that have systematically weakened the natural defences of the plateau, set high in the Western Ghats. Satellite images clearly show that Wayanad has at least 48 stone quarries, all run illegally, by the hand-in-glove arrangement, the corrupt businessmen-state nexus. The fact that the quarries are visible in the Satellite images suggest that they were either operational or abandoned without the required back-filling of pits and plantation trees, which is mandated under the law to ensure stability of the affected landscape.

The Nefarious Activities in Ecologically Sensitive Areas

World Weather Attribution, a global consortium of climate scientists who study the impact of global warming on extreme weather events, also highlighted in its August 14, 2024 report that "factors such as quarrying for building materials" have definitely contributed to the increased susceptibility of the slopes to landslides during heavy rainfall. It is also important to note that illegal quarrying generates a lot of black money—a perennial financial menace in Kerala.

To understand the spread of quarries in the Wayanad district, which ranks 13th of the 147 landslide vulnerable districts in India, as per the Landslide Atlas of India, the Satellite imagery of quarrying sites from open-source database OpenStreet Map was analyzed. The imagery was then superimposed on a map of the district to assess if the quarries fall in ecologically sensitive areas (ESAS) identified by the High Level Working Group (HLWG) on the subject in 2013.

HLWG was established by the Union environment ministry in April 2012 to formulate an "all-round and holistic approach to sustainable and equitable development" for the Western Ghats. This Group was led by Dr. Kasturirangan, former chairman of the Indian Space Research Organisation (ISRO). The group identified 4156 villages, spanning 59,940 km^2—37% of the mountain range—as "ecologically sensitive" and recommended a strict ban on quarrying, in these areas. These villages had at least 20% of the area under natural landscape, including regions with high biological diversity, wildlife corridors or heritage sites. Accepting the recommendations, the Union environment ministry in November 2013 issued directives under section 5 of the Environment Protection Act, 1986 which prohibits quarrying, mining and other environmentally damaging activities in ESAS. The very surprising and distressing fact is that 15 of the 48 quarries in Wayanad district are located in the 13 villages identified as ESAS. In fact, 9 of the quarries are within forests in two ESA villages, namely, Kidanganad and Noolpuzha.

Explosives' use in quarrying generates vibrations which disturb the soil profile, which is particularly dangerous for Wayanad's critical slopes prone to landslides. Quarrying not only creates fractures in the rock but also widens existing ones, allowing water to very speedily seep through following heavy rainfall, which ultimately, leads to landslides, due to the loosened soil profiles.

A Protracted Battle

Even after a decade of establishing a legal framework to ban mining and quarrying in the Western Ghats, India is yet to finalize the boundaries of ecologically sensitive areas to implement it. The following is a chronological assessment of the various initiatives:

1. August 31, 2011: The Western Ghats Ecology Panel submits a report to protect Western Ghats. Recommend strict restrictions that are opposed by the Kerala Government.
2. October 29, 2013: Kerala Government sets up a three-member panel expert committee to study citizens' concerns on the HLWG report.
3. December 2013: Centre asks the 6 states of the Eastern Ghats to suggest modifications in the boundary of the ESA villages.
4. March 10, 2014: Centre issues first draft notification to finalize ESA boundaries, but, it lapses as states demand smaller area under ESA. More draft modifications released in 2015, 2017, 2018, and 2022, but, all lapsed.
5. August 17, 2012: The center sets up HLWG to find a balance between conservation and sustainable development. Report submitted on April 15, 2013.
6. November 13, 2013: Centre accepts HLWG report and issues directives under section 5 of the Environment (Protection) Act, 1986. Prohibits quarrying in villages and ESA as identified by the HLWG report.
7. December 3, 2018: Centre accepts Kerala Government's demands and amends EPA, brings down ESA area spread from 13,108 to 9993 km^2.
8. July 31, 2024: Centre releases draft notification after the Wayanad landslides to finalize the ESA boundaries!

What Is in Store for the Future?

More Wayanad—like disasters could be in the making across Western Ghats as none of the states that share the mountain range have finalized the ESAS even though a decade has passed since the Union Environment Ministry accepted the Kasturirangan report and amended the ESA to provide a legal framework.

Kerala currently has 110 quarry permits awarded up to 12 years and 444 quarry leases awarded up to 3 years according to the Department of Mining and Geology. 11 of the 14 districts in Kerala have ESAS as identified by Kasturirangan report. There now 1688 quarry sites across Kerala, with 90 of them spread across 123 ESA villages. These quarries exist at a time when the state has not only weakened its regulations, but has, miserably failed to implement its mining laws. A 2017 Comptroller and Auditor General (CAG) report found that five Kerala districts, including Wayanad, lacked a system to ensure compliance by quarries. Of the 27 quarries inspected, 21 violated rules. The July 30 landslides simply proves that there is blood on the LDF government's hands.

Early Warning Systems (EWS)

Early warning system is an adaptive measure for climate change, using integrated communication systems to help communities prepare for hazardous climate-related events. A successful EWS saves lives and jobs, land and infrastructures and supports long-term sustainability. Early warning systems will assist public officials and administrators in their planning, saving money in the long run and protecting economies. The UN (United Nations), working in diverse partnerships, has introduced a number of innovative early warning systems initiatives in vulnerable areas around the world.

UNDP's (United Nations Development Program) "Signature Program" "Strengthening Climate Information and Early Warning Systems for Climate Resilient Development and adaptation to climate change" is a comprehensive program operating across Africa, Asia and the Pacific. Systems in place at the sub-regional and regional levels ensure preparedness and rapid response to natural disasters, using a model that integrates the components of risk knowledge, monitoring and predicting, dissemination of information and response to warnings.

In Uganda, where 64% of the population relies on agriculture for subsistence, most of the weather stations fell into disrepair following civil war, changes in government and poor maintenance. The Strengthening Climate Information and Early Warning Systems (SCIEWS) project has replaced outdated and inadequate meteorological stations with updated systems, improving disaster risk reduction with more effective means of generating and disseminating information.

In Southeast Asia, long coastlines and low-lying countries make the region particularly vulnerable to increasing extreme weather conditions. Kerala State is one such example. In Cambodia, for example, more than 1.7 million people in 2013 were affected by floods with a loss of $365 million; by 2016 the number of people affected by floods rose to more than 2.5 million people, indicative of an increase in climate-related flooding. A four-year program implemented by UNDP with the Government and other partners, entailed installing and re-activating existing Automatic Weather and Agrometeorological Stations and Automatic Hydrological Stations across the country. Farmers now can access climate bulletins for detailed information and carry out substantial planning to avoid costly consequences, both in money and lives.

UNEP's (United Nations Environmental Program) CLIMWARN Project, a partnership with entities in Burkina Faso, Ghana and Kenya, has replaced rudimentary meteorological methods with a more sophisticated system using modern technology that alerts communities of potential floods and other risks, which is then communicated to rural regions through SMS and email, aiding them in better preservation of their crops and livelihoods.

UN entities also provide support as implementing partners for the Climate Risk and Early Warning Systems (CREWS) initiative, which was launched at the UN Climate Change Conference in Paris in 2015. CREWS boosts capacity of Multi-Hazard Early Warning Systems and is operating in 19 countries in Africa and the Pacific, including Least Developed Countries (LDCs) and Small Island Developing States (SIDS), most prone to tropical cyclones and floods. In June 2018, CREWS announced that the program would be rolled out in the Democratic Republic of Congo, Burkina Faso, Mali and Niger. CREWS Steering Committee continues to identify new countries in Africa and Asia where programs can be financed.

It is heartening to note, though very belatedly, in September 2024, Kerala Government decided to instil EWS program in the State, to save live and property.

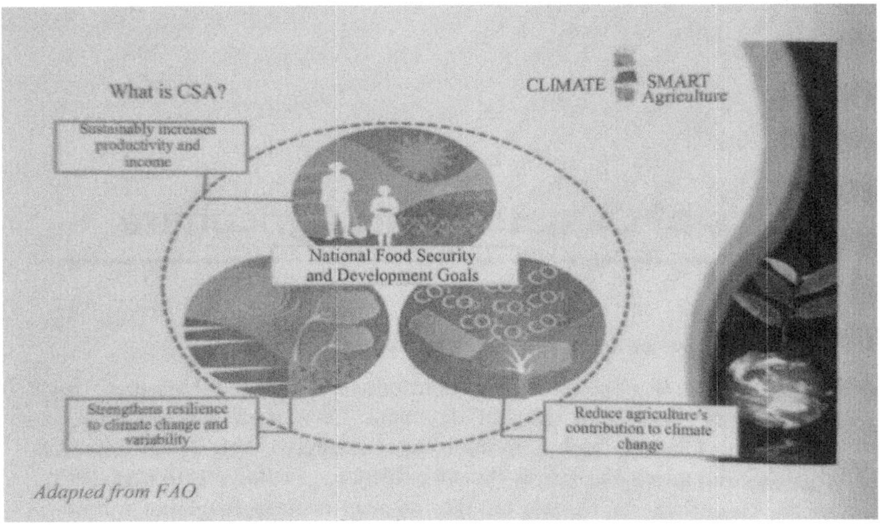

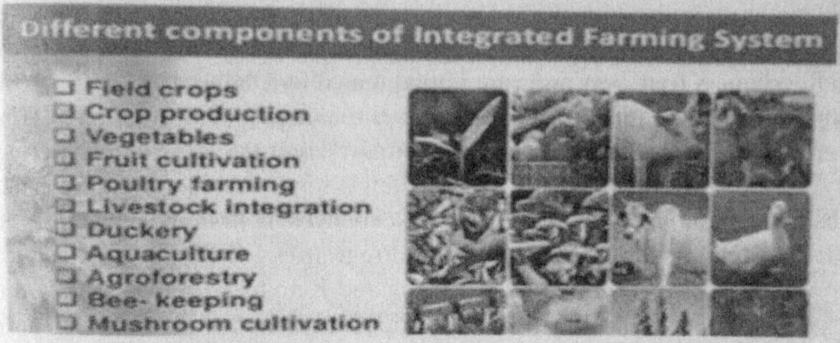

Fig. 9.16: Integrated farming system

Fig. 9.17: Solar irrigation system

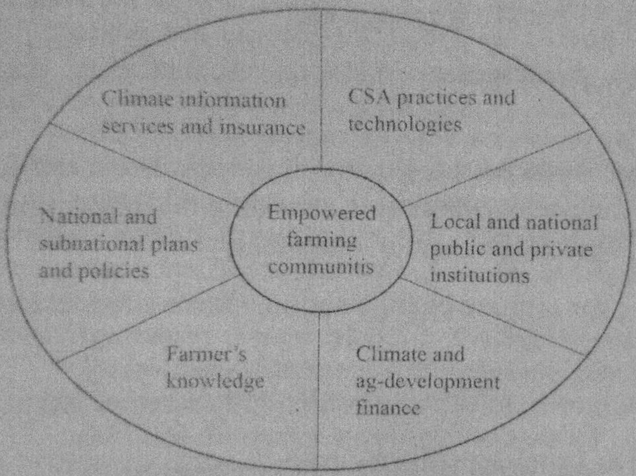

Fig. 10.1: Major components to be considered in CSV AR4D sites.

Fig. 10.3: Use of leaf colour chart

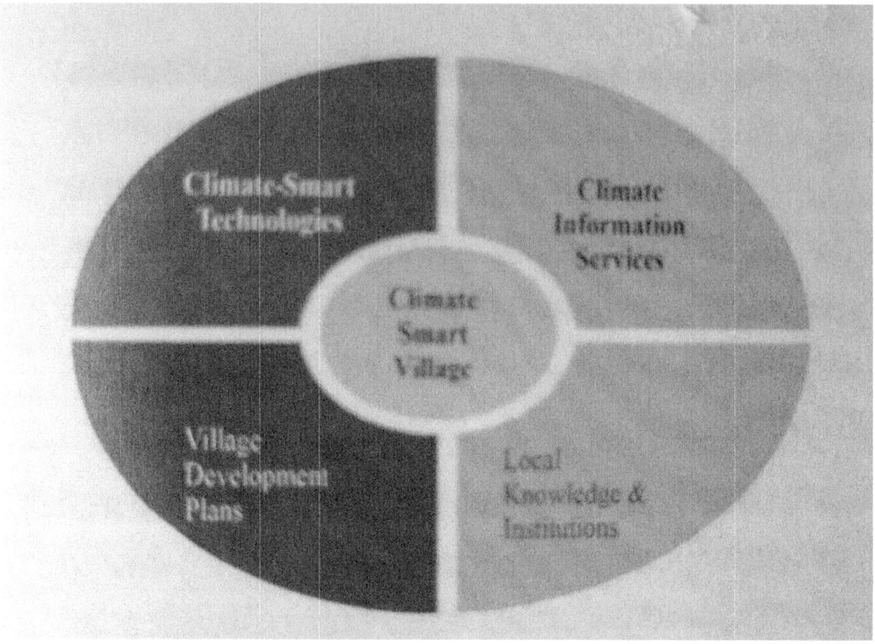

Fig. 10.7: Steps for setting up CSV (CGIAR)

References

FAO 2002. From farmer field school to community IPM (Integrated Pest Management) – Ten years of IPM training in Asia. Bangkok, Thailand (FAO), Rome. Italy

Franzel, S *et al.* 2015. Farmer-to-farmer extension. Good Practice Notes for Extension and Advisory Services. Lindau, Switzerland, Global Forum for Rural Advisory Services. Retrieved from http://www.g-fras.org/en/download.html

The manufacturer's authorised representative in the EU is Springer Nature Customer Service Centre GmbH, Europaplatz 3, 69115 Heidelberg, Germany. If you have any concerns regarding our products, please contact ProductSafety@springernature.com

Printed and bound by CPI Group (UK) Ltd, Croydon, CR0 4YY
01/10/2026
02225876-0005